ATLANTIC OCEAN
Ciego de Ávila
Camagüey
Las Tunas
Holguín
Granma
Santiago de Cuba
Guantánamo

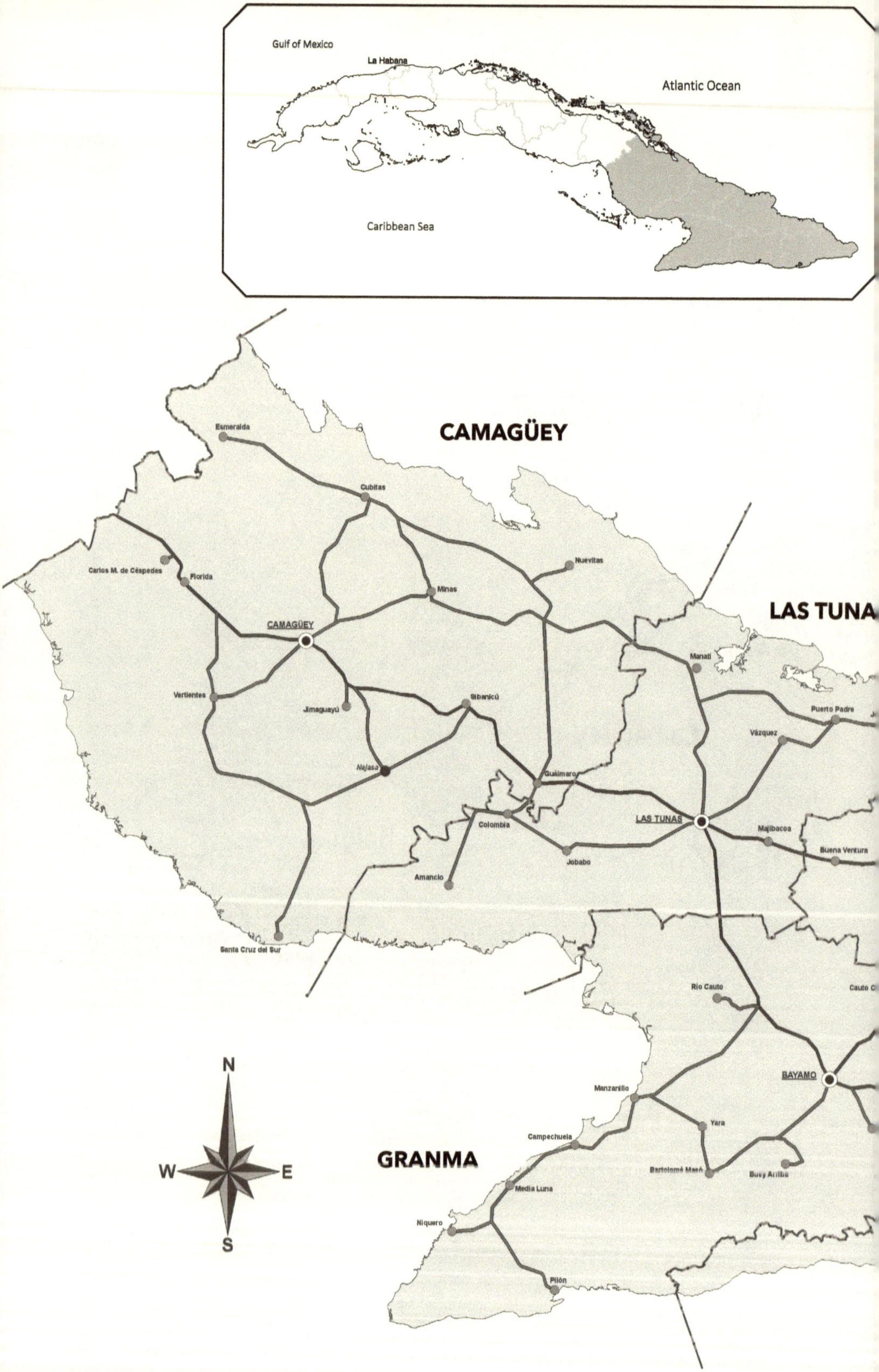
Gulf of Mexico
La Habana
Atlantic Ocean
Caribbean Sea
CAMAGÜEY
LAS TUNA
GRANMA
Esmeralda
Cubitas
Nuevitas
Carlos M. de Céspedes
Florida
Minas
CAMAGÜEY
Manatí
Vertientes
Jimaguayú
Sibanicú
Puerto Padre
Vázquez
Najasa
Guáimaro
Colombia
LAS TUNAS
Majibacoa
Buena Ventura
Jobabo
Amancio
Santa Cruz del Sur
Río Cauto
BAYAMO
Manzanillo
Yara
Campechuela
Bartolomé Masó
Buey Arriba
Media Luna
Niquero
Pilón
N
W
E
S

Communities and Settlements
Rojas Ramírez Grand Family

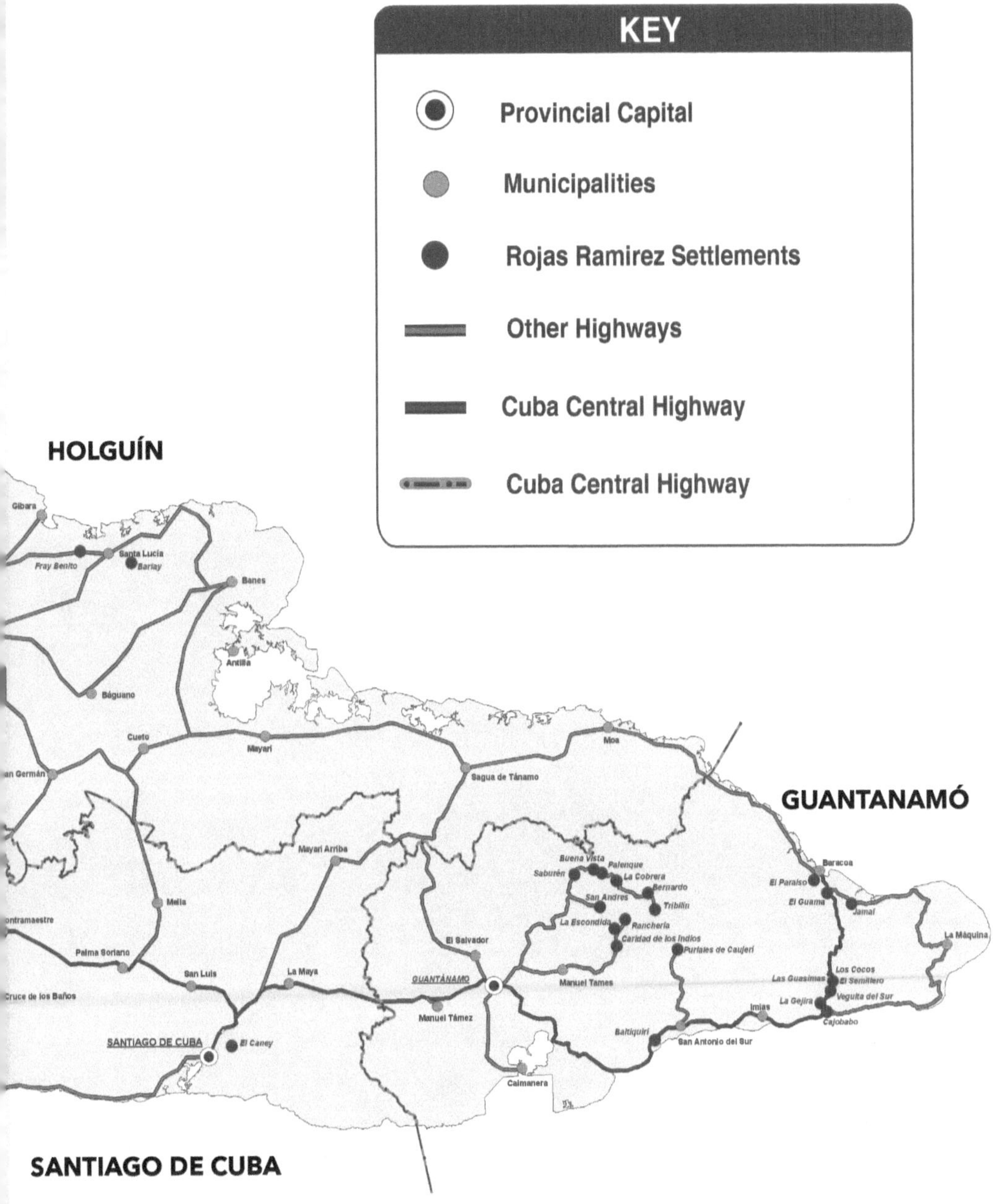

Dreaming Mother Earth

The Life and Wisdom of Native Cuban Cacique
Francisco "Panchito" Ramírez Rojas

José Barreiro

GUANÍ

Dreaming Mother Earth: The Life and Wisdom of Native Cuban Cacique Francisco "Panchito" Ramírez Rojas

Learn more online at www.guanipress.com

For more information regarding permission, write to:
info@guanipress.com
109 Echo Hill Rd
Saugerties, NY 12477
Photographs used by permission

Original Title: Panchito, Cacique de Montaña
© First Spanish Edition: 2001, Ediciones Catedral, Santiago de Cuba
ISBN: 959-7134-09-8

© Jose Barreiro, 2016, Second Spanish Edition: 2016, Fondo Editorial Casa de las Americas, Havana and Editorial Campana, New York
ISBN: 978-959-260-487-2

Design by Yolanda V. Fundora
Project Managing Editor: Mario Picayo

First English Edition originally published by Editorial Campana, 2018.
ISBN-13: 978-1-934370-73-5

Second English Edition, 2025.
ISBN-13- 978-1-962756-04-4

Library of Congress Cataloging-in-Publication Data is available on file.

9 8 7 6 5 4 3 2 1

Printed in the United States of America

Dedication

This book is for my father, Don Manuel Barreiro Cabrera (Camagüey, 1896–1981), guide of my days, who taught me where to search.
And to the memory of Ingrid Washinawatok el-Issa (Flying Eagle Woman), Menominee activist who loved Cuba, teacher of the mind-heart connection.

Acknowledgments

This first edition of the English translation of *Panchito: Mountain Cacique* owes deep appreciation to many people. I thank the teachers who inspired and helped this general work: José Juan Arrom and Manuel Rivero de la Calle, mind and body in the modern reencounter with Cuba's Indigenous roots; my colleague and coresearcher, Alejandro Hartmann Matos, historian of Baracoa and expert on the Indian legacy of eastern Cuba; Ingrid Washinawatok el-Issa, who inspired and assisted the Indigenous Legacies of the Caribbean Encounters; the researcher Isabel Lautín, who shared her sociocultural work with the Guamá Group and the people of La Caridad de los Indios; Antonio Martínez Fuentes, for his generous foreword and consistent attention to Cuban indigeneity; Ángel Graña (General Coordinator) with the Foundation for Nature and Humanity, for opening its research archives; and Nidia Pérez de Hartmann, who is especially appreciated for her eye on editorial detail and for the warmth of her family home. Appreciation is extended to Oscar Montoto, director of Ediciones Catedral in Santiago de Cuba, who brought out the first (Spanish language) 2001 edition of the book. Heartfelt appreciation is also due to Tim Johnson, Mohawk colleague from Six Nations (Ohsweken, Ontario), who consistently encouraged its publication; to Mario Llorente Leyva (Guantánamo), who partly translated the manuscript from the Spanish; and to Gabriela McBee, who conducted a second revision. (The final Spanish-to-English translation is fully my responsibility.) For a timely grant to assist with research and pre-production, we thank the stalwart commitment of Onaway Trust (England). For their familial embrace, and their generosity of mind and spirit, I am, of course, especially thankful to

Panchito, Reina, Idalis, Almeida, Vladis, Leonel, all the Ramírez Rojas *yucayeque*. And, for her personal and cultural embrace—in family and in mutual struggle for the life of the good—special thanks to my life companion, my wife, Katsi Cook. It is this circle that makes the present work possible. "*Hahom*," would say our Taíno ancestors, the natural people of Cuba. "Thank you."[1]

1 Taíno: ethnic group from the Arawak linguistic family. Proceeding from the Orinoco River watershed (today Venezuela), they migrated by sea in successive waves toward the Greater Antilles. An Amazonian base with influences from earlier migrations via Florida and Central America generated and amalgamated a mosaic of diverse cultural dynamics. I utilize the term Taíno to identify this predominant agro-ceramist culture of the Greater Antilles as indicated cultural root of the eastern Cuban region, one of the tribal nations in the Caribbean mosaic. L. A. Curet and others argue against the use of *Taíno* as group identifier for any native group in the Caribbean, reasoning that such use of the term to denote a people is not evidenced. Nevertheless, the related terms *taíno* and *nitaíno* are found to share a linguistic root in the Arawakan language family that consistently indicates kinship, blood, and family. Documented in use for nearly 200 years in Caribbeanist literature as indicative of the cultural traits common to the Greater Antilles, archeologists, linguists, historians, and ethnographers have repositioned the term over time. Contemporaneously, the term Taíno has begun to encompass the reemergence of Indio-identified groups across the island nations. We opt here to retain the use of the voice, "Taíno," as a powerful word with widely understood concept and for its poetic-historic-linguistic force. Caution is prescribed in considering proposals to discard its use.

Contents

Introduction

Knowing Panchito

The first time I saw Panchito was in a heavy rain; he came riding his horse down a mountain trail. At the edge of the square (*batey*) that marks his remote hamlet, Panchito—short and wiry—dropped sprightly from the horse and looped the reins on a tree branch. Leisurely through the dense rainfall he walked toward the family *bohío* (hut), welcomed by his wife and children, nephews, nieces, and grandchildren. His daughter, Idalis, our guide to the distant thatch-roof home, formally introduced us. We were in eastern Cuba, in a mountain pueblo, Caridad de los Indios, high up the Guantánamo mountains, and more precisely at the community of La Ranchería, the home of Panchito Ramírez Rojas, subject and testimonial voice of this book.

It was 1995. I was visiting Cuba to take part in the conference "Cuba: Nation and Migration" at a moment when the "special period in time of peace" seemed to be turning a corner on the severe economic deprivation caused in Cuba by the downfall of the Soviet Union. I took the opportunity to rent a car in Havana, and at the end of the conference to motor east to the plains of Camagüey, and then along the coast to Baracoa, and finally up the mountain to Panchito's home at La Ranchería.[2]

A small group of friends and relatives—piloted by Alejandro Hartmann, historian of the city of Baracoa and guided by Idalis Ramírez and local educator Isabel Lautín—clambered up the steep mountain on an old Soviet jeep, from the sugar town of Manuel Tames to the central village of the Rojas-Ramírez families, to greet the people of most distinct Indigenous genealogical, historical, and oral tradition in Cuba.

2 José Barreiro, "The Cacique's Prayer," *Native Americas* (Cornell University, Akwe:kon Press., March 1996) describes this incident. In this early North American recognition of contemporary Indocuban survivance, Panchito comments, "Our grandparents taught us songs, they showed us sacred foods ... you'd be amazed what we have here." Ricardo Gonzalez Padrón, president of Camagüey-Madison Sister Cities Association, also accompanied the mission.

That revealing and somewhat dangerous journey presented this writer with wonderful news. It was true, and I would ascertain for myself, the presence of the *gens* (clan) of the numerous Ramírez and Rojas families (later, other surnames would affix) of the Guantánamo-Baracoa mountains, who hold substantial oral and genealogical tradition stemming from, as principal source, our Indigenous Cuban roots.

Panchito Ramírez Rojas, sincere elder of this *sui generis* community, is a modest guajiro yet a human being of considerable depth. A natural philosopher, Panchito is, through his communal culture and by historical conviction, a Cuban patriot; he is, furthermore, a Cuban Indian cacique, having inherited the title from his grandfather (Cacique Ladislao), who carried it for two-thirds of the twentieth century.[3] That first night on the mountain, in his rustic dirt-floored bohío, we began what would become a continuing communication. That night's conversation went full circle until the following morning, as heavy rain caused flash floods and the danger of travel was undeniable. It was fortuitous, as Panchito himself declared, that "you had to spend the night in the mountain."

As a child growing up in the nearby province of Camagüey, I had heard from my aunts and especially from my father that, in the mountains inland from Baracoa, I would find hamlets of guajiro families of a deeply rooted Cuban Indian extraction. My father knew about it because during his own childhood, in 1906 at ten years of age, he took part in a guajiro healing pilgrimage to La Virgen de la Caridad del Cobre. It was a horseback prayer journey for his grandfather who was injured by a splinter to the eye while cutting wood. The story of that journey was particularly special for my aunts because the Copper Virgin granted their grandfather what they would hail as a miracle. Kneeling before the little statue of La Virgen del Cobre, after some two weeks on horseback and nearly moribund, Don Joaquín Cabrera,

3 Ladislao Ramírez Rojas, 1879–1972, great-uncle of Panchito Ramírez. Veteran mambí of the War of Independence, Ladislao appears in the list of combatants of the Hatuey Regiment, in Mayor General Carlos Roloff Mialofsky, *Índice Alfabético y Definiciones del Ejército Libertador de Cuba*, data compiled and ordered by the Inspector General del Ejército Libertador (La Habana: Imprenta de Rambla y Bouza, 1901): 746. More than eighty Rojas and Ramírez ancestors appear in the ranks of the combative Hatuey Regiment, notably at the Battle of Sao del Indio (August 1895). Cacique Ladislao was photographed by Rivero de la Calle in 1964.

mambí (rebel) veteran of the War of Independence and guajiro from Camagüey, begged and cried for a cure to his painful ailment. As he wiped the tears of his prayer with his red war handkerchief, out came the stuck splinter.

At the chapel of El Cobre, my father said, the guajiro pilgrims from Camagüey met up with Indian families from Oriente who were there to pay back a promise made to the Virgin during a birth that had gone well. As an old man, in 1978, he repeated the story for me. "The origin of what is guajiro, the culture of the *guajiro del monte*, comes to us from there, from the mountains. If you want to meet the most pure *gente india* (Indian gens) in Cuba, go to the mountains between Guantánamo and Baracoa. That is where the Cuban guajiro comes from."[4]

In the early 1980s, news articles began to appear in Cuban publications (*Granma* and *Verde Olivo)* reporting on families from La Caridad de los Indios and nearby mountain areas. I would come across such articles at Olin Library at Cornell University, when US Customs allowed them to enter the country. The infrequent articles brought minimal information, sometimes by young journalists with inane notions, but also some detailed intriguing reports about research from those mountains by senior Cuban anthropologist Dr. Manuel Rivero de la Calle. These reports, illustrated by photographs, supported the continued existence of the mountain population that my father had related.

Mutual work during those days with Caribbeanist maestro, Don José Juan Arrom, enlivened my search. With Arrom at the Smithsonian National Museum of the American Indian to videotape on Caribbean mythology, we conversed on the exquisite archeological statue of Deminán, Creator twin of the Taíno Antilles, central to a Caribbean cosmology. "Go to the eastern *sierras*, young man," the maestro encouraged me. "If I were younger I would go with you. You will find ethnographic treasures there. Those are our roots."[5]

We spent the night with Panchito that first time. It was a night of rainstorm, thunder, and near tragedy. In a flash flood produced by

4 Hayde Moya León, Granma National Newspaper, May 9th, 1996, p 5, titles her article on Panchito Ramírez, "La Caridad de los Indios: A Sui Generis Community."

5 Ramón Pané, *Antigüedades de los Indios*, ed. José Juan Arrom (México: Siglo XXI, 1975). In the search for Taíno knowledge and roots, Arrom is an indispensable maestro. The quote is from personal correspondence with Barreiro. See "Arrom: Caribbean Creation" (video documentary), Smithsonian National Museum of the American Indian, 1996.

heavy rains, the swollen watershed of a small brook suddenly hit and washed away our old Russian jeep. As the jeep began to float, we were saved by young men of the mountain, who made a human chain to pull us out of the raging river.

Later, in the tender embrace of a nearby Indian bohío, in dry borrowed clothes, warming ourselves by the fire, each of us with our plate of rice and beans, we talked long into the night. And we confirmed something exquisite: Panchito has deep expression, rich with natural, generational wisdom; based on his own sui generis history, he carries on his own native intelligence.[6] Sitting at the open door of his bohío, surrounded by family on the top of a Cuban mountain, Panchito Ramírez Rojas asserted: "We are Indians. This we carry through the generations. We are Cuban Indians from these mountains. To me, my grandfather said: 'Grandson, what we have left of our Indian culture should never fall. Keep being who you are.' That is why I repeat in all certainty that there was no extinction here. No, the Indian is here, rooted. And as long as there is a generation of mine, I will not allow it to fall."

It is in great measure thanks to the work of Dr. Manuel Rivero de la Calle that we know who Panchito is and who were and are his people. Rivero de la Calle studied the Indocuban population of the eastern mountains during the 1960s and 1970s.[7] He centered his attention on Manuel Tames municipality, Guantánamo, where La Caridad de los Indios is located. Using methods of physical anthropology, Rivero categorically asserted what was already retained by oral tradition and had been reported in the 1940s in articles by Antonio Núñez Jiménez and by a somewhat obscured yet significant chain of historians and other observers: There are enclaves of Indocuban descendants in the eastern mountains, a gens of numerous families, many with the surnames of Ramírez and Rojas, along with other surnames, located in the Guantánamo-Baracoa area and elsewhere in the Sierra Maestra. These clans of extended families have consistently claimed their Indian identity; sometimes applying the term *indio*, as well as the term *natural*.

6 Hayde Moya León, *Granma*, May 9, 1996. On p. 5, León titles her article on Panchito Ramírez, "La Caridad de los Indios: A Sui Generis Community."

7 Panchito generally used the term *Indio*, but on this first visit he also referred to *Taíno culture*. Apparently his use of the term *Taíno* results from general information contact through the 1970s and is not likely a mountain traditional reference.

Rivero de la Calle conducted exhaustive research over two decades, involving blood analysis, measurement of dentures, size of head, lips, arms, and other traits and features. He and his collaborators assert something appreciable, that Panchito and his folks are indeed Cuban Indians, clearly exhibiting all the physical characteristics associated with the Indigenous peoples of South American and Caribbean origin: "Short in height, scarce hair growth, straight, very black hair kept into very old age . . . internal and external eye fold (epicanthus) . . . moderate obliqueness in the eyes, light brown skin with a reddish tendency, wide noses . . . "[8]

Rivero worked his rigorous study in two phases, working intensely in the communities of Yateras, Caridad de los Indios, and La Escondida and reviewing preliminarily other places in the extensive region of Guantánamo to the Bayamo environs. The anthropometric study proved valuable. Although he does not focus on oral history and tradition, he notes of the Rojas-Ramírez families, that "they are extended over a great geographic area, almost all carrying the surnames of Rojas and Ramírez, demonstrating that these are surnames of great antiquity in the zone."[9] The respected Cuban anthropologist points out that the Rojas-Ramírez folk are "one of the largest families in Cuba, I believe the largest, without having to deepen the subject."[10]

Many people will be surprised that at the beginning of the twenty-first century, four hundred years after their supposed extinction, bases of the Cuban ancestral root are still asserted within the insular Cuban population. Yet there are clear signs that the thesis of extinction of the Indian ethnicity and genealogy in Cuba and in the Caribbean generally continues to reveal itself. The persistence of native legacy is not limited to the homesteads and small *caseríos* (small villages) of native descendants, such as the Rojas-Ramírez gens; it extends notably to what might be considered an appreciable current in the national ethos.

---

8 Ramón Dacal Mouré, Manuel Rivero de la Calle, *Arqueología Aborigen de Cuba* (Editorial Gente Nueva, 1978): 157. Dacal and Rivero, writing about only La Caridad de los Indios, find that "... more than a thousand people from this area present such characteristics . . . known as 'Indians,' no one doubts who they are."

9 Manuel Rivero de la Calle, "Los indios cubanos de Yateras," *Revista Habana* (1973): 151–172. Rivero cites Gates on the origin of the last names *Rojas* and *Ramírez* in mass baptisms celebrated at the time of the Spanish New Laws in the mid-sixteenth century.

10 Manuel Rivero de la Calle, in a letter to Pedro Soto, November 12, 1973, Rivero de la Calle archives, copy provided to Dr. Antonio J. Martínez Fuentes, University of Havana.

The Cuban population has a notable Native American biological inheritance in apparent extension of the original people of the island and of the Caribbean. Recent studies on the national genome indicate that 34.5% of Cubans inherit Native American mitochondrial DNA. The highest levels are found in the eastern region provinces of Holguin (59) and Las Tunas (58), confirming a previous study giving a 33% estimate.[11] This is comparable to the genetic work realized in Puerto Rico by Dr. Juan Martinez Cruzado, where the high index of MtDNA also demonstrates high Indigenous incidence in the maternal lines within the population—the Caribbean phenomenon we tend to call our Indigenous Matria or Matrix.[12] Maria Nelsa Trincado reminds us that "For nearly two centuries [the sixteenth and seventeenth centuries], albeit declining, the Aboriginal population was in good measure the majority and it is evident that it prevailed in many aspects of Spanish and Black daily life...." She adds that: "still, during the eighteenth century in Bayamo, Santiago and other towns, neighbors complained that, during the night, they heard drums and songs of the *areítos* [Indigenous ceremonies] announcing nostalgia, possible uprisings or mass suicide the next day...."[13]

Notable scholars such as Miguel Rodríguez Ferrer (1847), Adolfo Bastián (1875), Stewart Culin (1900), Mark Harrington (1915), R. Ruggles Gates (1954) among the most distinguished, mentioned and sometimes traced brief profiles of this particular Indocuban group. In the case of the Rojas-Ramírez, the gens descends from the "pueblos of

11 Beatriz Marcheco-Teruel, Esteban J. Parra, Evelyn Fuentes-Smith, Antonio Salas, Henriette N. Buttenschon. "Cuba: Exploring the History of Admixture and the Genetic Basis of Pigmentation Using Autosomal and Uniparental Markers." *Plos Genetics* (July 2014), available at http://www.plosgenetics.org/article/info%3Adoi%2F10.1371%2Fjournal.pgen.1004488. Also, Isabel Mendizabal, Karla Sandoval, Gemma Berniell-Lee, Francesc Calafell, Antonio Salas, Antonio Martínez-Fuentes, David Comas, "Genetic origin, admixture, and asymmetry in maternal and paternal human lineages in Cuba." *BMC Evolutionary Biology* 8(1) (July 2008). DOI: 10.1186/1471-2148-8-213.

12 Juan Martínez Cruzado, "The Use of Mitochondrial DNA to Discover Pre-Columbia Migrations to the Caribbean: Results for Puerto Rico and Expectations for the Dominican Republic," *KACIKE: The Journal of Caribbean Amerindian History and Anthropology*, special issue (2002), ed. Lynne Guitar, available athttp://www.kacicke.org/MatrtinezEnglish.pdf.

13 María Nelsa Trincado, "El aborigen y la formación de la nacionalidad cubana," *Anuario El Caribe Arqueológico*. (La Habana: Casa del Caribe, 1996): 102+.

Indians" established by Spanish policy of "La Experiencia" in the mid-1500s and applied in the eastern region during the colonial period.[14]

Crossing the Valley of San Andres de Yateras, in March 1847, Rodriguez Ferrer observed a community called "Place of the Indians" comprised of nine families of Rojas-Ramírez, for a total of 122 people. In detailing their genealogy he writes, "Here is a trunk of this tribe." He meets the "District Lieutenant," or headman, Don Juan Rojas y Aranda. Rodriguez does not record the honorific of "cacique," but describes Rojas' work as chief and as "justice of the peace" of the community. Rodriguez notes that the people make "much of their pure Indian race." He adds that they reclaim being Indians, and "of those from the conquest."[15]

Culin (1902), half a century later, notes that "in Dos Brazos, between Yateras and Baracoa, there are one hundred families of Indians, called Rojas-Ramírez. They live under a cacique named Juan Anguita."[16]

In 1945, the geographer Antonio Núñez Jiménez, exploring the eastern mountains, reports the presence of an Indocuban population, both on the coast south of Pico Turquino and in the mountain area of Yateras. "The Indians of Yateras, who likely total some three thousand, have left for themselves only the refuge of these mountain forests....It is curious that almost all carry the surnames of Rojas and Ramírez."[17]

Núñez Jiménez gives us a population estimate of three thousand Indigenous rooted Cubans in Yateras, and perhaps more in the region of Pico Turquino. Ramón Dacal Mouré (co-author) and Rivero

14 For works that illuminate the aboriginal history and Indian pueblos in Cuba, see Hortensia Pichardo, "Los Orígenes de Jiguaní," Facetas de Nuestra Historia (Santiago de Cuba: Editorial Oriente, 1989): 77–100; Olga Portuondo, "Una sublevación de indios en 1758," *Revista de la Biblioteca Nacional José Martí,* no. 1 (January–April 1981) La Habana: 199–204; Felipe Pichardo Moya, *Los indios de Cuba en sus tiempos históricos* (La Habana: Editorial Siglo XX, 1945). As well, the excellent essay by Juan Manuel Reyes Cardero, "La Inserción del Aborigen en la Sociedad Colonial Santiaguera: El Caso del Pueblo Indio de San Luis de los Caneyes," *Ciencia,* no. 1 (2009): 3–16, available at www.redalyc.org/articulo.oa?id=181321570001.

15 Miguel Rodríguez Ferrer, *Naturaleza y civilización de la grandiosa isla de Cuba* 1 (Madrid: España, 1878): 455+.

16 Stewart Culin, "The Indians of Cuba," *Bulletin of the Free Museum of Science and Art of the University of Pennsylvania* 3(no. 4, May 1902): 209+.

17 Antonio Núñez Jiménez, *Cuba: con la mochila al hombro* (La Habana: Ediciones Unión Reportaje, 1963): 197. Núñez describes his expedition through the Toa River in 1945. Núñez relates his face-off with a cacique of the time, Celestino Rojas, who had recently run off from the mountain a young engineer who had pretended to measure away their land. In "El Pico Turquino, Exploración y Estudio" (La Habana: Sociedad Espeleológica de Cuba, 1945): 37, Núñez writes about the coastline of the Turquino region: "These are found almost entirely inhabited by the descendants of Cuban Indians, from whom they conserve part of their primitivism."

calculate for Caridad de los Indios that "more than a thousand persons in this region present these characteristics…they are known by the name of Indians and nobody doubts who these people are." Rivero further affirmed that not only are Indian families residing in Caridad de los Indios. but that there are other nuclei of such families in other areas of the Sierra Maestra and in the eastern provinces generally.

There are today some seventy or more caseríos of multiple Indocuban families, what Panchito calls *lugares de familia* (family places), in the environs of Caridad de los Indios, with others also well-established along the Cuban mountains toward the coast of Cajobabo and Baitiquiri, west along the southeastern ridges, to the valleys of Bayamo, and to the plains and coasts of Camagüey and other diasporic points. A family count requested by the families is in progress, with the assistance of regional museums. So far it approximates Núñez Jiménez's numbers but is quite incomplete and anticipated to yield six thousand or more. As the family-by-family count proceeds, the sharing of family information and stories tends to cohere the extended communities of the larger kinship group.

In the mountains of the Nipe-Sagua-Baracoa massif, communities and *barrios* (neighborhoods) of the Rojas-Ramírez gens can be encountered in the small towns of La Ranchería, La Escondida, El Ácano, Los Pozos, El Pinar de Ceiba, Vega del Cedro, Vega de Los Toros, and La Caridad. Toward the valleys of Yateras, extensions of these families are found in the following villages: La Alegría, Arenal, Arroyo Bueno, El Cilindro, Bernardo, Jucaral, Purialito, Felicidad, Palenque, Tribilín, San Andrés, La Cristalina, Santa Cornelia and Porvenir. New fieldwork in the area continues to identify Rojas-Ramírez villages in the Valle del Caujerí, Baitiquirí, and also in San Antonio, Puriales, Imías, Veguitas, Las Guásimas, Cajobabo, Baracoa and Maisí, Holguín, and Bariay. On the regional and national level, a diaspora of small nuclear groups of the grand family extends into Maisí, Guantánamo and Santiago de Cuba, Las Tunas, Holguín, Camagüey, Matanzas, La Habana, and Isla de la Juventud.

La Ranchería, identity centrifuge of the *Gran Familia* (Big Family), is a community of profound presence on its mountain, such that it has merited the recognition of "autochthonous community" from the Province of Guantánamo and the Municipality of Manuel Tames.

The assignation as *Comunidad Autóctona* is based on earned respect from the state for the self-governance evidenced in the family-based authority of Cacique Panchito Ramírez, and in acknowledgment that the community is repository of an efficient model of food self-sufficiency and exemplary for its knowledge of traditional agriculture and its example of social peace, with almost no index of theft, crime, or violence.

There are men and women elders in many of these families who retain considerable oral tradition passed by the generations, on various ecosystemic production systems, with detailed attention to natural forces, healing properties of plants, conuco and other cropping methods, knowledge of domestic modes of production, accompanied by ceremonial practice, with cultural-spiritual orations that request strength from natural elements.[18] Beyond physical appearance or suspect notions of "racial purity," this natural intelligence is what truly engages the context of this work. Enthusiasm sustains for direct knowledge of nature among native cultures—Indigenous, guajira, *mestiza*, *criolla*—because they tend to rely on humanistic and collaborative values, especially in the agricultural and healing practices of everyday use. In Panchito's expression can be noted points of millennial observation. If his beliefs assimilate a blend of cultures, the core of his traditions are clearly Indigenous American and resonate with other known aboriginal lifeways. This is evidenced in Panchito's clearly expressed spiritual perception and his attention to the interchange of energy with nature, in his prayer that recognizes the natural cycles of continuous creation, in his burning of *copal* (incense), cedar, and tobacco as sacraments (smudging), in the connection with the ancestors and with the Creator force in its cosmic dimension.

Among Panchito's expressions, many relate generally to American Indian cultures: the prayer to the four directions, or celestial points; the system of asking permission from plants before taking them as medicine (reciprocity); the desire to keep the conuco (companion-planting) cropping system; the system of planting and tree harvesting by the phases of the moon; the use of dreams as guides for spiritual

18 See José Antonio García Molina, Daisy Fariñas, Mercedes Garrido, *Huellas vivas del Indocubano* (La Habana: Editorial de Ciencias Sociales, 2007): 30–31. The authors gather testimonies detailing persistence of Indocuban cultural expressions in the spiritist ceremony "Dance of the Chord" and other contemporary Cuban customs.

healing. It is worth noting that these are knowledge bases with Indigenous ecological value. Most revealing of indigeneity is Panchito and the elders' expression of tender affection toward the widespread Amerindian Cosmic Family complex: Mother Earth; Father/Grandfather Sun; Grandmother Moon; the four directions or four winds; and for the sacred elements of fire, air, and water. Panchito speaks humbly yet eloquently on these and other matters, much of which finds convergence in Hispanocuban and Afrocuban expressions, not for that reason diminishing in substance nor form as expressions of indigeneity within cubanity.

Clearly, Panchito and his contemporary generation and the many and extended Rojas-Ramírez families represent a genuine kinship with the population of Indocubans known by such eminent figures as José Martí and Antonio and José Maceo in the nineteenth century and studied by Antonio Núñez Jiménez and Manuel Rivero de la Calle, among others, in the twentieth century. Historically consistent, everything indicates the representation of a deep ancestry rooted in the Cuban archipelago. Rivero de la Calle and Núñez Jiménez agree with the sustained assertions of Indigenous legacy by the contemporary population. The historical testimony of Rodriguez Ferrer corresponds, in March 1847, when he tells us that the Rojas-Ramírez Indian community in San Andres "... make much of their origins in the pure race, of those of the conquest." In respectable continuity we can see that, some 168 years ago, Panchito's direct ancestors proudly proclaimed their identity as Indigenous people of Cuba.[19]

Some folks get intellectually entangled when considering efforts to reconceptualize the Indigenous legacy of an American country. For Cuba and the Caribbean in general, it has been a remote and minimized topic. But those who study the American cultures always discover that indigeneity as a current of identification persists among the peoples and cultures of the Caribbean. To be sure, it is a complexed and nuanced identity and this presentation is not at all an example of cultural or racial purity, nor defined to invite or raise issues about "internal nations" within such a politically articulated and culturally integrated island-nation as Cuba. Panchito's people and other enclaves

19 Miguel Rodríguez Ferrer, *Naturaleza y civilización de la grandiosa isla de Cuba* 1 (Madrid: España, 1878): 455+. Rodriguez Ferrer faults the Indians for primitive agricultural methods while noting their general food self-sufficiency.

of Indian descendants identified in this work are not differentiated populations, nor primitive forest tribes, untouched by colonization and so-called modernity.

For Cuba and beyond, the word of Panchito Ramírez impresses with its sincerity and simple profundity. Panchito's voice is equally valid in its Indigenous legacy as in its expression of the knowledge of the forest and the culture of the Cuban mountain guajiro. This important ecosystemic wisdom of the mountain cultures—"the Indian science," as Panchito calls it—has been maligned in the advance of Western science. At the dawn of the twenty-first century, however, the eco-agricultural knowledge of domestic mode of production is seriously contemplated as concept of adaptation (mitigation) to new climactic conditions. This evidences in the two currents that braid Panchito's thinking: his indigeneity and his cubanity. If a large current of his natural world knowledge has extended through the Cuban farming culture and practices in general, not for that is his ecosystemic knowledge less Indigenous. Nor, because it may have blended with other supposed races, is the Cuban gens less autochthonous.

Panchito Ramírez and his community, along with all the other enclaves of Indian families, are a people of the contemporary historical moment. Their children go to school. All, including the elders, feel quite Cuban, in the broadest sense. They express clearly that they love and will defend the Cuban nation. In their cultural discourse, they identify with the Revolution, with which they share history, while simultaneously they maintain a sense of roots in the land that is both millennial and post-historical. As in all sectors of Cuban society, in the present generation they self-represent in a range of professions, including medicine, education, law enforcement, and engineering.

No one can doubt there have been great changes since the time of the conquest, including the blending of many peoples. Yet out-marriage (*mestizaje*) does not decide the extinction of a people any more than having a certain DNA necessarily decides ethnicity. Nor should the indigeneity marker of people be determined in the freeze-the-culture "ethnographic present" approach of an inherited, thoroughly colonial anthropology. Our focus shifts rather to consider the family relations, tracking the gens or related trunks of families of strong biological and cultural inheritance, and most essentially, the everyday intentionality of maintaining a common transgenerational memory, a living social current with a sense of conscious adaptation.

A valuable, authentic expression of the guajiro culture of the Cuban mountains, Panchito says of himself, "I am a *cruza'o*," meaning, of blended traditions. Panchito is a healer and a farmer—he is a musician who drums the Indian beat of the mountain but can also perform the traditional *sones* of the Changüí music of Guantánamo, as well as African songs of *bembé* shared by the earliest encounters—and he is Cacique, or elder of conscience, in his community.

Cacique, an authority

The persistence of a community-recognized leader titled "cacique," in a population with a profound sense of kinship and continuity of existence, is in itself a strong negation of the myth of Indocuban extinction. It is truly impressive that persons such as Panchito—individuals who balance contemporary society while connecting to deep cultural roots —are still nurtured and valued within the Cuban people in general. No doubt all nations that can celebrate the persistence of culture and the continuity of identity will grow.

Within that experience we can contemplate the survival of a community concept of authority, continuous for over a century, that honors and uplifts an elder as cacique. We have the pleasing personal experience of observing over twenty years the specific leadership of Francisco Ramírez Rojas, of being privy to the recognition of a cacique, as carried by Don Panchito, and granted upon him by not only his community of La Ranchería but by many of his extended gens in the region and other places nationally.

The consistency of a community authority titled "cacique" in Caridad de los Indios was noted by Culin (1902).[20] While in Baracoa, Culin wrote: "The President of the Club [Club Unión] told me that there are around a hundred families of Indians called Rojas and Ramírez in Dos Brazos, between Yateras and Baracoa. They live under a *cacique* by the name of Juan Anguita." The Guantánamo historian, José Sanchez Guerra, in his documentation of the case of the Catalan

20 Stewart Culin, "Central American and Cuban expedition (1901)," field notes, Stewart Culin Papers, University of Pennsylvania Museum of Archaeology and Anthropology, Archives of Philadelphia, May 1901; Stewart Culin, "The Indians of Cuba." *Bulletin of the Free Museum of Science and Art of the University of Pennsylvania* 3 (no. 4, May 1902): 185+.

midwife who recruited the Yateras Indian warriors to the cause of Cuban independence in 1895, reports the existence of several caciques, major and minor ones, among the Indocuban population of Caridad de los Indios.[21]

Rivero de la Calle got to interview the elderly cacique Ladislao Ramírez (1879–1974), Independence War veteran, mambí, and a great-uncle of Panchito. Addressed by his people as "Ladislao Cacique," the elder led the mountain Indian community of Rojas Ramírez for over seven decades of the twentieth century.[22] Community tradition about Ladislao Cacique holds that he would visit among his people's farms daily, leading land purchases and apparently trying to improve the conditions of Indian relations. Ladislao Cacique also took on the responsibility of paying salaries of teachers for the new generation of his people.[23]

In 1969, at the age of 97 (some claim he was 107), Ladislao Cacique was interviewed by the national newspaper *Granma*. His testimony was brief, but very curious, with historical detail still in the common memory. The elder recounts his participation in the War of 1895, as well describing his community's migration to the mountains from "the hills of El Caney and Yerba de Guinea, near Santiago de Cuba." Among other topics, the elder Ladislao gifted the reporter a transcendent cultural memory: "[Ladislao] mentions that he came to know legends transmitted by his ancestors who worshiped idols built of stone. 'My grandparents believed in the Sun and Earth,' he affirmed."[24]

The term *cacique* was reported by Columbus and is noteworthy for its early adaptation from the Taíno language. It denotes at La Ranchería "an elder or community leader of greatest respect." The tradition of selecting a cacique appears as inherent custom within the jurisdiction of the first historical Indian pueblos, such as El Caney and Jiguaní, both

21 José Sánchez Guerra, "La Capitana del Regimiento Hatuey," *Revista El Mar y La Montaña*, (Guantánamo, 1998).

22 Ladislao Ramírez, 1879–1974, great-uncle of Panchito Ramírez. Veteran of the War of Independence, he appears listed among participant members of the Hatuey Regiment, See, Mayor General Carlos Roloff Mialofsky, *Índice Alfabético y Definiciones del Ejército Libertador de Cuba*, data compiled and ordered by the Inspector General del Ejército Libertador (La Habana: Imprenta de Rambla y Bouza, 1901): 746+. A count of eighty-one Ramírez-Rojas ancestors, just from Yateras, appear in the ranks of the combative Hatuey Regiment, whose Indian troop fought with special valor in the Battle of Sao del Indio, among other combat encounters in the War of 1895. Cacique Ladislao was photographed with Rivero de la Calle in 1964.

23 Francisco Ramírez, recorded testimony, *CILP Archivo*.

24 Joaquín Oramas, "La Escondida: ¿un refugio de los taínos cubanos?" *Granma*, November 30, 1969: 9+.

of which titled their main community leaders as caciques. "That thing of cacique," as Panchito says, with characteristic modesty, "is natural for us Indians. The cacique is a kind of local judge."[25]

The elder Opublio Rojas Ramírez (1920–2009) uttered the term while greeting us in our first visit to La Ranchería. "The Cacique Panchito will be here shortly," he said, after informing us that, in fact, "you have arrived among the Indians."

The authority of Panchito, his condition as venerated elder, is immediately recognizable. We met him on a day of rain, in his bohío at La Ranchería, surrounded by children, nephews and nieces, and grandchildren. From the first moment we witnessed a respect that has been constant for more than twenty years.

A sui generis cultural process occurs in the transition from Cacique Ladislao to his successor, Panchito. The group intentionality to support the protagonism of its leader in the new generation is evidenced in numerous details and consistent attitudes.

Cacique Ladislao led the Rojas-Ramírez gens families from his return to Caridad de los Indios after the War of Independence (1895–98) to his death in 1974. The phenomenon is well documented; Rivero de la Calle interviewed and photographed the old cacique. Ladislao was great-uncle to Panchito and oriented his young "grandson." Panchito recounts: "…[Ladislao] as an old man would speak a lot to me and people could see that. Because here people take note of the young children, then and now. My grandfather [great-uncle], Cacique Ladislao, would tell me: 'Look, the one who will carry our Indian culture forward is you, grandson, it has to be you.' Since I was a boy he would speak to me that way.

"...When the grandfather [great-uncle] died, the people here told me, 'Panchito, Ladislao who has cacique has died, then too, the elders Emiterio and Yayimín, both died, that leaves you or your uncle Opublio.'

"Opublio, who is older than me, did not accept to be called "cacique." He would not accept. So he sends it…[the title]…to me."

The leadership tradition and the personal charisma of Panchito braid inexorably. Elder, leader, and profoundly knowledgeable farmer

25 José Barreiro, "Panchito: Cacique de Montaña," 2nd ed. (La Habana: Casa de las Américas, 2016).

is also a healer, peace judge, and primary spokesman. He conducts the principal ceremonies of the community.

"...I don't like that they called me 'cacique,' because I don't want to be confused with a bossy type. Myself, I would prefer they simply call me 'Panchito.' But my people put that on me, that of Cacique Panchito, so what am I to do. Because we are Indians, and that of calling one elder person 'cacique,' that's natural around here...that of cacique came from before. That was to be the principal grandfather, or family judge. Before Ladislao, there were other caciques...so people titled me that, as cacique, perhaps because I am interested in everything that might be considered as our culture."[26] That interest for the mountain knowledge of his ancestry, what this leader calls "the Indian culture," is a generational commitment, a promise to the grandparents and elders. Says Panchito: "My grandfather [great-uncle] said to me, 'Grandson, the Indian tradition should not fall. Keep up who you are.' That is why I repeat with every certainty, here we had no extinction. Here the Indian is planted. And as long as there is a generation of mine, I won't let it fall."

It is notable that Panchito describes cacique leadership as a "local judge," in a term comparable to the one noted by Rodríguez Ferrer's 1847 visit to the Rojas-Ramírezes at the Indian Estancia. Declares Panchito: "That of cacique came to us from before. That of being the principal elder or family judge." And, of additional relevance: "... because that's the way I was raised by my grandparents, my uncles, my parents, and even my older brothers, from whom I learned a lot. That is why I have always been for what is good. That strengthens me as cacique."

Through several decades Cacique Panchito has enlisted assistance in bringing his Indocuban community to public recognition, to "let ourselves be known" and thus express their culture. From the first years of his time as cacique, since 1974, Panchito organized members of his community to participate in cultural events in the region. In the past twenty years, his group attended events in Baracoa, Santiago, Camagüey, and Havana, with the intention of counter-arresting the historical condition of invisibility, revealing the lie in the myth of Indigenous disappearance.

---

26 José Barreiro, Panchito, Mountain Cacique: A Guajiro-Taíno Testimony from Francisco Ramírez Rojas (Santiago de Cuba: Catedral Ediciones, 2001).

Panchito has protagonized a renewed respect for the mountain Indian culture. In his expression, he gathers the notable agro-ecological practice of the elders and many young farmers. Traditionally, he values self-sufficient food production, teaching from a strong dose of traditional ecosystemic knowledge, extensive use of medicinal plants, and all wrapped in a genuine cultural memory, the *imaginario* (fabulosity) of the Cuban guajiro.

Panchito recruited us at our first visit in 1995 to his strategy of correlating key points of their community oral history with the academic archives. Over time, we would be able to confirm with good historical documentation his claim ("on behalf of the *ancianos*") that not only Ladislao, but an entire contingent of Indians from Yateras, Caridad de los Indios, and other parts fought in the Independence Army under General Antonio Maceo. The cacique intended to have a studied response to the exaggerated rumor and historical trope of a collective "treason" by Indians in that war. In new research and archival documents, we were able to find the revolutionary Hatuey Regiment, with over eighty Rojas-Ramírez combatants in its ranks, and evidence of heroic and bloody engagements in Cuba's favor by this regiment.[27]

The sorry state of their housing and the lack of electricity was an issue of long-term angst for the community. In the remote mountains it is not always possible to obtain building tools and materials. The cacique petitioned strongly for better housing options and at one point a house was offered to him with the promise that other houses would be constructed "in coming years." The cacique refused the offer with words that have become legend: "And I, am I a man and my people are devils? No, that cannot be. Let's wait for a solution for all. Either we are all humans or we are all devils." It took another five years of civic and prayerful requests, but the new houses were finally constructed, and with solar panels as well.

A few months later, the Indian mountain community celebrated the housing achievement at La Ranchería. A long evening ceremony—an *Altar de Cruz*—was offered in spiritual payment. A ceremony of blended customs, the *Altares* are celebrated in song and prayer. On this occasion, the Altar is brought to La Ranchería by godparents from the nearby community of La Escondida and other caseríos. This is a

27 José Barreiro, "El Regimiento Hatuey: más allá del mito de la extinción," *Herencia Cubana* 12 (no. 3, October 2006): 46–51.

necessary duality of the ceremony. The godparents of the ceremony blessed all the new houses with water at the four main corners, and they celebrated the community leader. Among Indians, the Altar de Cruz is celebrated with a distinction: to the patriotic and the Christian elements are added the effigies of the sun and the moon. An unforgettable song gains for a singer from La Escondida the prize of the small Cuban flag. His song intoned: "The tradition of the Indians cannot fall," thus "we sing to Panchito, the only cacique in Cuba."[28]

Panchito, elder of the small Indocuban community at La Caridad de los Indios, obviously a man of our times, nevertheless keeps a spiritual connection with our most essential Cuban and, consequently, most natural of American traditions. In line with one custom that has persisted, his folks call him "cacique," a term which, more than rank, reflects affection and abiding respect.

In the years 1997 to 2003, Panchito directed field workshops at the scientific-cultural conferences, "Indigenous Legacies of the Caribbean," held in Baracoa. Panchito—as well as Doña Reina and other Taíno elders—was the heart and soul of these cultural and academic encounters. People from different countries came to talk with him, to listen to the music of his mountain relations, and to accompany him in his herbal walkabout along the Toa River.

Since then, always in the company of numbers of his people, Panchito has visited communities of native descendants throughout Cuba (Guirito, Jiguani, Niquero, Camagüey, Havana), and has extended the recognition of his mountain Indian people to various events.[29] Significantly, in 2002, he and twenty of his people were invited to represent the Cuban Indian people and tradition at the International Habano Festival. In 2003, another major step, he and the elder women community leaders officiated in the repatriation and reinternment of Taíno human remains by accord of the Smithsonian Institution and the Cuban government.

Year by year from 2004 to 2013, Panchito and his groups of elders and youth have been active representing their community, history, and traditions at many places and venues in Cuba. Significantly, in

28 See "La Rancheria: Mito de Extinción" (video documentary), Armando Guerra & Osvaldo Capote, directors, Onaway Trust, 2010: www.youtube.com/watch?v=OJ0phUUhqZw.

29 José Barreiro, "A Bridge for the Journey: Trajectory of the Indigenous Legacies of the Caribbean Encounters, 1997–2003," *Indigenous Resurgence in the Caribbean* by Maximilian Forte (New York: Peter Lang Publishers, 2006): 235–252.

August 2014, the cacique and daughter, Idalis, his "walking cane," (*bastón*) represented the Cuban Indian community at an international conference of Indigenous Peoples, sponsored by Casa de las Americas, in Havana.

The narration in chapters that follow this introduction is Panchito's direct testimony, the words of a Cuban Indian elder, a descendant of the eastern Cuban Taínos. It was Panchito's own decision to tell it. Mostly candidly, sometimes guardedly, he describes a world of natural thinking presented here with deep respect. The life and voice presented here are entirely his own, characteristic of his natural expression, which includes access to more formal education only briefly and as a mature man. Panchito is essentially nonliterate.

My work as editor/transcriber of Panchito's words has been to recognize the obvious, to interview, record, and type, and to read and work with him to certify his precise meanings. I helped him take his manuscript to print, not only to provide added evidence of the significant existence of Indigenous roots among the Cuban people (as well as about the extent of indigeneity in Cuban culture), but most importantly because from within a profound Cuban identity (*cubanía*), Panchito's testimony reveals an intrinsic natural humility about an actual and true way of life.

Getting to know Panchito has been a spiritual gift. Across the gulf that divides the Cuban nation, may we all agree to greet him as the wise, honest Cuban that he is. The Cuban people, in general, have their own spark, a genial idiosyncrasy; and the Cuban guajiro people, still today, have their native perspicacity. Panchito is all of that. Thank you, Cuba, for what you do well. Thank you, José Martí, for seeing it all. Thank you, natural man of Cuba, for your generous sincerity.

José Barreiro

Akwesasne Mohawk Nation, 2015

Being a cacique is not about commanding. My authority, if I have any, comes from letting people know me, letting people love me. And that, I feel, is what I'm doing here with you, with these words. I believe I'm letting people get to know me.

Life

That of being cacique

I prefer to be known as Panchito, simply that. I am from here, from the mountain. I have never moved. You find me on the land, in the conuco, in the little farm where I was born.

It's true that sometimes they call me "cacique."[30] And I want to explain that so there's no confusion. They say it's my turn to be cacique, my own people say it; the old ones put that on me, and even the scientists who have been coming here in recent years from Cuba and from abroad have called me so.

But it's not that I ask for it. No one asks to be cacique. Being a cacique is not about commanding. My authority, if I have any, comes from letting people know me, letting people love me. And that, I feel, is what I'm doing here with you, with these words. I believe I'm letting people get to know me.

I start with this: Here in La Ranchería, in our community, we do the good. And I've always been for what is good, because that's how I was raised by my grandparents, my aunts and uncles, my parents, and even my older brothers and sisters, from whom I learned a lot. This reinforces being called "cacique." So for me, I prefer to be called Panchito. Panchito Ramírez Rojas, a Cuban guajiro Indio from La Ranchería, from here, from the mountain of Caridad de los Indios.

But I'm not saying we didn't have caciques of the Indios before. The term *cacique* was used here and that's what it was called. One generation ago the eldest here, my grandfather or [actually] great-uncle Ladislao, was called Ladislao Cacique. That grandfather of mine fought in the War against Spain; he was a mambí. People respected him for his intelligence. He didn't read, but his farm produced for him and he brought a teacher for his children and grandchildren here in the mountain. As an old man he talked a lot to me and people saw it. Because here, in the past and now, people pay attention to children. My grandfather [great-uncle], Cacique Ladislao, would say, "Look, the one who's going to carry the Indian culture is you, grandson, is you."

30 Cacique: 1:a native Indian chief in the Caribbean and areas dominated primarily by a Spanish culture 2: a local political boss in Spain and Latin America./ https://www.merriam-webster.com/dictionary/cacique

He talked to me like that since I was little. And you see me here where I am still with this, for me, tremendous culture.[31]

So people chose me as cacique, because I'm interested in everything that comes from our culture. It's not that I'm a bossy cacique. Perhaps, it is more that I'm one who loves his people. When the grandfather [great-uncle] died people said to me, "Panchito, Ladislao, who was cacique, died, then Emiterio and then Yayimín, who also was cacique, died. Now we have you and Opublio." And I said, "No, but I'm Panchito, not a cacique." Imagine, Opublio who is older than me, doesn't want to be called cacique, either; he won't accept it. So he passes it onto me. And, as I say, I don't like it either, since I don't want anyone to mistake me for a bossy person. But that's what they call me, Cacique Panchito; what can I do? Because we are Indios, calling the elder "cacique" is common around here.

Later on, scientists came from abroad. They started coming and always asked that question, "Where is the cacique?"

Research was done here in the seventies, when some Soviet professors came to study the Indios. That study amused us. They measured our teeth, our jawbone, our hair, our height; they checked our blood, our cheekbones; well, so many measurements only to show what we already knew. "Let's see, are they Indios or not?" Because they'd extract blood, measure the size of our mouths, our noses; everything they measured completely. And to us it seemed strange, since from our grandparents we already knew who we were. But the study confirmed what had been so much denied, that after all we are people from La Escondida de Indios and that we are Indios.

During their work with us, the scientists witnessed Cacique Ladislao's death.[32] That was the loss of our guide, our most natural elder. That's when folks started looking for me more and more. Because someone has to do things, for the birth of a child, or holding a wake for the dead, anything. And the scientists noticed that this is how it

31 Joaquín Oramas, "La Escondida: ¿un refugio de los tainos cubanos?" *Granma* (1969): 9+. In 1969, at around ninety years old, Cacique Ladislao was interviewed by the national Cuban newspaper, *Granma*. His testimony was brief but curious, with historical detail still in the common memory. The old man tells of his participation in the War of 1895, and he also describes the migration of his own ancestors from "the mountains of El Caney and Yerba de Guinea, near Santiago de Cuba." The jurisdiction of El Caney as an Indian town always had its cacique. Ladislao gifts reporter Oramas with a transcendent cultural memory, insisting "that he got to learn the legends transmitted by his ancestors, who adored idols constructed with stones. *My grandparents believed in the sun and the moon*, he affirms."

32 Panchito is referring here to Dr. Rivero de la Calle and colleagues, on expeditions in the early 1970s.

was. "Well, Cacique Ladislao died, now it looks like you are left as the Indios' cacique," they said as well.

But as I say, the thing of cacique came to us from the past. It meant being main grandfather or family judge. Before Ladislao there were caciques also. I knew what being a cacique meant for our people. But I asked, "And to you, what does it mean to be cacique?" They said, "A chief, respected, respected in the tribe." I said, "Well, I have to get together with all my people to prove it. Since I alone can't decide it, because the word alone doesn't make you a cacique; you aren't a cacique just by name."

That's how I thought it should be, if I were to be called cacique. Others who have studied more should tell me if I'm wrong or not. The cacique is cacique because he knows about everything in his community. He's the one to know where his place and his authority are rooted, which is mainly in one's morale, in one's honor. It's with this understanding then that people from our mountain call me "cacique" and accept my view.

That's what I have to offer as cacique and that's what makes me cacique, so it has nothing to do with legal but with moral authority. If it were up to me, I'd like them to call me, rather than "chief" or "cacique," I'd like to be known as "Panchito." It would be nicer to be known as "Panchito," the one who carries on with the Indian culture. My way is: I'm respectful and I want respect for myself, my humanity. What makes me a cacique is that I'm not abusive with anyone. I've never spanked any of my children, never.

Certainly, among people being cacique has some rank. A cacique can pull someone's ears who won't listen, because when I intervene in a party brawl in the mountain, people say, "Look, the cacique is here." This means I can say to a troublemaker coming to the fiesta, "Do you want me to tie you up to the tree I have for tying up troublemakers?" And I can do it and I can have him there for the night tied up, punished. I say, "I'll let you go in the morning when the sun rises."

I want peace, not violence. Folks see that and perhaps that's why they respect and like me. I've never had to punish anyone for real, because they obey me through reasoning. I talk with people. I tell them we're all humans and being so we should help one another, enjoy ourselves at a party and at a Mass or at an Altar of the Cross, celebrate the spirits, the saints as well as Mother Earth, as Father Sun, who are also sacred.

So the term *cacique* was already being used before the scientists. The cacique has to be like a sun for his community. My grandfather [great-uncle] was like that. And they had given that name of cacique to my grandfather Ladislao in his generation with formality, since he was the eldest in the area and since he was a mambí who fought in the war along with Maceo.[33]

Now, on our mountain, when someone dies they come to where I am and say, "Look, we need you to organize this." And then I'll say, "Let's go, this needs to be done." Everyone responds to the call I make, whether it is for someone who died or for a sick person. Or for a party. They grant me that respect.

So at our place the authorities don't have to come and watch over a party; we ourselves resolve things with my authority. I tell them, "Alright boys, anyone getting rude here, we already know where we'll put him." And people say, "Anything, Panchito. Let us know."

What I do ask them is not to call me "cacique" so much, just "Panchito." But respect, for sure. My example, if I see a young man going too far, I call him and say, "Look, you're young, you have to learn from us old ones about respect." At a party, if I see someone going too far, I say, "Hold on. Listen, this is respect."

For instance, I don't drink rum, I don't smoke, I'm not rude to anyone. So, when I'm going to talk to a person, I do it with that authority and popularity. And with the required respect people respond.

My folks are Indian

My folks are Indian, they always were, here at La Ranchería, here at La Escondida, the Indios' hideout, which is what La Caridad de los Indios was. Here the elders were always respected. And I belong to my people here, to the Ramírez and the Rojas and the other families with Indian roots around here in these eastern mountains.

Although we have been here very long, few people know us. That's why they say we Cuban Indios don't exist anymore, but the truth is we

33 Antonio Maceo (1845–1896), Mayor General of the Cuban Liberation Army, also known as the "Bronze Titan" of the Cuban wars of independence.

do, we're here. Yes, it's true that we're from the Cuban Indian generation from the Oriente (East); we are here, in Yateras, in Maisí, in Jiguaní.

In Cuba we're all even. There's no discrimination, neither against blacks, nor against whites, nor any mix. But I tell to whoever listens that the Indio exists, we exist, we are Cubans and we are Indios. Through our grandparents we know that we are Indios from here, from those people who are our ancestors. Certainly, we are guajiros from the mountain, very Cuban from the time of Taínos. And well entrenched to defend our homeland, united forever with Cuba.

You ask me if there was extinction. I say to you, *no*. My parents said the same. And I repeat, "They killed many; many died, but the Indio is here, rooted." My grandfather said to me, "My son, don't drop our culture."

And that's why since I was a child I concentrated on knowing the things that have made it to us, these things that come to us from former times, from this, our natural life in the mountains, and from what we remember of the ancestors' teachings. More than anything else, it's because I remember my elders' ways that people come to see me. I think about these things and people talk with me. I have families who come to ask for my advice on life. Sometimes other Indios ask me, "Why don't you come over here to share our culture of the mountains with us?" And I say, "We'll go, we'll sing, we'll share ceremonies to the plants and to the trees, to the earth and the sun." Because in these mountains, there are families and hamlets; there's plentiful land here of aboriginal people.

Sometimes a leader approaches me to ask, "How are you doing, Panchito, how's your health?" Because captains, lieutenants, generals come here, and those who care about my health hug me and some say, "Follow your culture and never leave it behind; follow it." And that's what I want, that's what I want, of course without offending my country and being proud of it, because my country taught my children, because all my children can read and write, and for that I give thanks to the Revolution.

Origin of my love for the mountain

What I tell you comes to me from my childhood, from nature, from my grandfather, the one named Juan Rojas Ramírez and whom we called Cancio, and from my grandmother Lina, who both told my mother to take good care of me, that I was going to be sensitive because they saw the clear light in me and because I didn't quarrel with other kids, but instead took care of them, I loved them. If you ask me, "Panchito, where does your medicine come from?" I tell you that's where it comes from. It comes from my childhood, my dreams and things, my tremblings. And also from my men and women saying to me, "He's from the good side." They opened me on that side.

I never quarreled with any child. When I saw other children fighting, I separated them and told them not to, because we were playing. There always are children who are aggressive but I would say "no" to them. My grandmother cheered for me and my grandfather said, "Oh my, this one, this one is going to be clear." They blessed me and patted me on the head.

They gave me lots of advice. "Seek to treat the people well," my grandmother used to say. "That no one cause trouble among us." And he'd say, "I was a fighter, a fighter with Maceo and Martí, and you have to be the same. Cuba must be freed." That's how my grandfather spoke to me and I said, "I will, grandfather. When I'm a man, I'll defend my country just like you." And I think that's what I'm doing, defending my country.

Because Cuba has its troubles. And I, Panchito Ramírez Rojas, farmer from the mountain, want to see if through my culture I can help my homeland and my government. I plant, I love my crops, but not only by sowing corn, sowing beans, can I help, but also by seeking our beautiful Indian culture, which is one more thing to help our country.

I want my folks to be recognized, and that they recognize themselves as being, some more and some less, natural people from here. Yes, we have marriages with others, but the root remains here. So they continue liking the mountain, loving the countryside, the beautiful nature we have, because the mountain is where the real thing is, it's where everything comes from. And we, the people of the mountain, have our value.

I believe the Indigenous in our *costumbre* [tradition] is going to be respected more and more, because it's a very Cuban thing. Look, I love the mountain. I don't like living in the city. My whole life I've belonged to the mountain. I love the freshness, the air that flows there, the crystalline water we drink, where all is natural. The fruit we pick is natural; we don't have to cool it. The tomatoes we pick, the beans we eat fresh, the fresh meat, and all that is life. For human beings, that's the most beautiful there is. I find that, as a human being, what my culture asks from me is to want to live from nature, to live from the earth, without taking anything from anyone else.

And I have my community. I'm not saying it's a developed community, yet it's true that we live like the five fingers of one hand, knowing how to pull together. We all aren't the same, but we're all equal because within all of us runs blood. Over in Caridad are ten houses and if a pig is butchered all ten houses take for free and I believe that's what our country is looking for, that thing of community, that cooperation.

And that way of being also comes to us from forever, the way of the Indio; if we made *casabe* [cassava bread], everyone ate; if we hunted, everyone ate; if we fished, we all ate fish and we traded it for casabe and for the food we produced. When they hear about Taíno, to me that's the Indian custom. What I believe was Taíno, it's the conuco, families helping each other.

That's how we grew up. I was raised with that humanitarian spirit that all Cubans should have, but unfortunately not all of us do. But I do have it; I carry it in my mind. Unfortunately I can't give the feeling to everyone. However, I do explain the problems to my community, I explain our values, how we must live as human beings. Because our grandparents used to say, "Let there be no fight among Indios." And you hardly see a fight here. If there's a brawl between two neighbors, I make sure nothing happens. "There can't be violence here," I tell them. "Here we all have to be equal." Even on my farm I tell them, "Alright, let's plant because the soil wants to be planted." Because that's what gives us life.

I am a farmer, I plant; that's my greatest pride. When I see that beautiful tree over there and the grasses swaying in the wind, for me it's the finest, most delightful thing there is. I sit down under a mango tree, there on a rock, and watch the bean field over there, moving with the air and I enjoy myself with that, watching it dance, watching the grass, watching the yuca, watching all edible things. Then I stand under an avocado tree and I say to it, "How beautiful nature is, and how delicious what it offers so that humans may live!"

Certainly, humans have to plant and cultivate it, clean it the way Mother Earth told me. I'll tell you about that later. Because I feel very full from that message, from when in a dream, the Mother of the Earth spoke to me.

Reality of dreams

Sometimes I dream I'm plowing through vegetable gardens with my oxen and planting. In the morning I tell my children, "Listen, last night I dreamed I was planting beans, or I was planting sweet potatoes or yuca," and we have to go plant, because when I dream it, it means I have to do it. And then the young ones say, "Alright old man, let's plant."[34]

They call me "*viejo*" [old man]. "Old man, come on, what day do you want to plant beans?" I always tell them, plant on a Wednesday afternoon, because you plant beans on a Wednesday and when you're going to pick them, you pick them very, very clean, without a single leaf, only with the pod. All the leaves fall off the stalk.

That's a dream system I carry. For instance, I see in a dream that an ox is going to attack or that it's coming after me or after any other person. I tell the young ones, "I dreamed last night that an ox attacked," and then they say, or my wife tells me, "Well, that's how it happened." Because it's true that many of my dreams materialize. While sleeping it seems that I'm seeing that animal and spirit tell me that it's going to attack or it's ready to do something violent. So that's the faith I have in my dreams, because many things in my dreams have turned out to be real. They're messages they bring to me.

You didn't know it, but before you had done it, I dreamed of two women dressed in white, and when the group actually sent a white dress for my daughter, I said, "I saw it, this was what I dreamed about, that is the cloth the woman in my dream had, bright white." So when I left the house I said, "When I come back, we need to do some work on that cloth, in the temple where one can see it." Look what clarity a dream gives you, especially the dream one remembers, and your spirit

34 The impulse or duty to "make reality" or to "comply" with dreams is widespread among Indigenous American cultures. See Michel Perrin, "The Way of the Dead Indians: Guajiro Myths and Symbols" (Austin: University of Texas, 1976): 7+. Here Perrin describes a dream tradition among the Wayuu or guajiro of Venezuela, another people from the Arawak trunk of Indigenous peoples: "If a Guajiro starts dreaming that he is elsewhere, near a well or in a house . . . that means his soul has left his heart, . . . to fly away over yonder . . ."

seems to go wherever you are, where you go in your dream, where you see it all. If I told you all I've seen! Because a spirit takes you to see things, to see the medicine that should be prepared. Spirit tells me, "You're going to see all the beautiful things there are everywhere we go." That way my dreams bring messages to me; they encourage me to carry things out. They tell me, "Look, this needs to be done. Watch this crop how tall it is."

The knowledge that we ought to carry out dreams into reality comes not only from my own feeling, but from my people as well, because my mother used to say, "Every time you dream something, try to make it real." And that's why she'd say, "My son, be watchful with your dreams."

As a young man I liked to always dress in white, bright, bright white, white pants, white shirt, and if not, a black shirt with white pants or black pants—straight colors, so there is no deceit. That used to be my wardrobe, but oh well, since our Cuba is subject to a blockade, now we have to dress with what we have. But my style was clarity, it was white, pure, pure white I wore, and it was clarity. I follow that conscientiously, too. When I see someone all dressed in white, from the Virgen de la Caridad del Cobre or from Santa Clara,[35] I like it, and on my own path I ask *El Mundo* [the World] to take care of me, to give me strength.[36]

My children and others in my family also dream. They have good dreams. They dream something is going to happen to a certain person when taking a particular path. And I tell them, "Well, pay attention to those dreams." Just last night, she [pointing to his daughter Idalis] dreamed that a cousin of hers had been killed back there because of a problem with a brother-in-law, and I told her, "Those dreams are hard to interpret; it could be just a worry one has." But sometimes it is a dream of warning, a message, and I pay attention to that, because that warning comes from the other side.

---

35 Santa Clara was born in Asís in 1193, and followed her fellow citizen Francisco de Asís in the path and vow of poverty, founding the order of nuns called the Clarisas. Her life was austere and rich in works of charity and piety. She died in 1253. See *Oficio Divino Liturgia de las horas según el Rito Romano IV. Tiempo ordinario – Semanas XVIII–XXXIV.* Liturgical co-editors (España, 1984): 1099+.

36 Panchito refers to "El Mundo" as a central spiritual entity, thus capitalized. To dress in white corresponds also with consecration in Santería, in the cult of San Lázaro, in Babalú. In the San Lázaro cult, as in other Cuban traditions, associations are noted with Taíno cemi (spiritual deities), such as Baibrama, Corocote and Buya and Aiba. See Daisy Fariñas Gutiérrez, *Religión en las Antillas* (La Habana: Editorial Academia, 1995: 125–128.

Dream of Mother Earth

I'm going to tell you a dream. It started a few years ago. But it's a vivid dream. Among my dreams, I believe it is the strongest I've had. I dream I'm at home sleeping in a hammock and suddenly I'm in the hammock but outside my house, on a hill over there at La Ranchería. In the sky I see a woman. She's in the white of the sky, in lilac color and she is an Indian woman, just like us, copper color, long, black hair, a very beautiful and very large woman. She shows a tenderness that calms me, and asks me in the dream, "Do you know who I am?"

I tell her that I don't know, that I only know that she is the grandest and most magnificent woman I had ever seen in my life.

She tells me, "I'm the Mother of the Earth, the mother of you all, not only of you here but the whole of the Earth."[37]

She was high up in the air and I gave myself to her in love. And I said, "Good, Mother Earth, and what would you have us do?"

Mother Earth said, "Plant on me, plant plenty of food, plant much corn, plant many conucos so you all can live. You need to protect everything there is on Earth, take care of the animals, take care of the human beings like you, don't burn the forests because you burn my arm, you burn my fingernails."

She said, "Human beings no longer suckle directly my breast. This makes me sad." She told me that now few people, and mostly only animals, still eat directly from her.

"Live from me," she said. "Suckle from my breasts."

All that she told me. I don't know if it's because I believe, because I'm religious, and [because] for me the true religion is she herself, the Mother of the Earth. It might be because I belong to that natural humanity, but she spent a while in that dream with me. She spoke about the buildings and the big cities and all those great things of human beings. "If they bother me," she said, "all I need to do is to twitch."

In that dream Mother Earth requested help; we should help her. We need to cultivate her a lot. She said we have to plead and request from her, love her, because she has so much to give. I also heard these words from her: "Ask the sun for me to purify the land, to heat up the

---

37 The concept of the Earth as mother is common to many Indigenous agricultural peoples of the Western Hemisphere. "Pachamama," say the Quechua; translating as "transforming energy-mother earth." "Etanoha," say the Onondaga of central New York, meaning "the Mother Earth," intoned within the Iroquois Thanksgiving Address.

soil, but that he not burn me, so that what you plant can grow, so that everything everyone plants can grow, so that everything the whole world plants can grow, every elder, every child, everyone belonging to humanity."

She was a *lila* woman, brown color, Indian color like us. That's how I saw her in my dream. It was a very clear dream and she was talking to me. I could see my body inside the dream. And I even pondered about my own things within that dream. What Mother Earth was saying seemed like an answer to me. Because I have indeed asked the Earth, I have prayed to her; I have cried to the Earth.

The truth is that there is something strange. There is a lot of drought. Here in the mountains you feel it. Water is getting scarce. I've thought there's a lack of love for the earth, there's no tenderness with little plants. What the ancestors used to say about loving the earth, people hardly keep that custom anymore. That's what I'd been thinking. In my Indio heart from here from the mountain, that's the pain I'm carrying.

Then that great woman says to me, "A human being from Earth is just a feather for me. I pick him up like this and he's a little feather, but he has a power, they have a power, because they make many beautiful things, many pretty things."

I told her then, "Mother Earth, take care of us."

She replied, "Yes, I take care of you, because you are my children, but you take care of the Earth."

And that's what I do. I plow the soil, but with pain, because I know I'm breaking her, wounding her. I have to do it, to be able to plant the crops I grow, because I plant in her, because I was born here on this land and I am a grower. I'm old and I haven't left this place, and I don't intend to leave this mountain, not until I disappear.

Childhood at La Ranchería

In my childhood I couldn't study because my father was very poor. Those times were ugly. In all of La Caridad de los Indios there was only one school, which was a little thing. And our community used to be much larger.

La Ranchería is a historic place because there were Indian camps, there were small Indio ranches here. For instance, La Escondida was their hideout. San José was like a hiding place in the mountain where the Indian people used to hide, and they lived there. Because at the beginning, when the Spaniards cursed the Indios, those ancestors of ours had to flee into exile; they had to abandon and search for shelter deep in the forest. We have La Lora, Bernardo, La Escondida, La Caridad de los Indios, and other places. These were places where Indios came to hide and form their bohíos and bateys. They called it La Caridad, my grandparents told me, and we call it La Caridad de los Indios, because there once was a woman of ours, a pure India, named like that, and because of her the name of the farm of La Caridad remained.

I tell folks about this little place of ours, because, well, who is it that still lives here? I say this place is historic, for the hideouts that were formed here. I hold this place as a thing of history. And it's my place; I was born where my grandfather lived, where my father lived, where my whole family lives now, here, this little historic place. I have faith in that knowledge because my grandfather said it to me, and he was a sincere man, one of those who knew how to say, "This is this," and that's it.

In La Ranchería itself, in my childhood, everyone was Indio—there was no other race and there were twenty-five houses, and today only eleven little houses remain. We were twenty-five family houses, aunts and uncles, cousins, brothers, and sisters. Very few other people would get there. We were poor people, but we were a community and we worked hard. Of course, when they could, children played; what we couldn't do was study in a school.

I liked to shoot with the sling; our grandparents used to do it, and I suppose those Indios from the past did, too. For the sling, you'd take a strip of *yagua* [palm] and with a stone we'd shoot from one top of a hill to the other. We were four or five youngsters and we'd compete to see who threw the farthest, and a stone thrown by one of those slings reaches far. My grandfather told me that Indios with slings used to overwhelm Spaniards. That was an Indian weapon, the sling. Listen, if it hit you in the temple it killed you. It knocked an eye out, broke a bone. A stone in a sling kills animals, birds, or hens, and we all played with it.

We played baseball, marbles, ball, spinning tops, all that since childhood. And we'd run a lot; since we didn't have schools to study in, that's what we did. I used to play a lot of baseball. The ball we made out of jagüey[38] because back then there was a government that handed out nothing but beatings, it was Batista's,[39] and to play we made the balls out of jagüey with rubber inside.

I used to pitch and hit well, and I often was given a ribbon. In those days you hit a home run and the girls came out with the ribbon. It was a nice custom. And when you won a game or were pitching well, they'd buy a candy for you, or they'd give you a soda or something. It was an incentive.

As children we used to do something like freestyle wrestling, where two men place themselves head-to-foot on the ground and lift one leg each, interlace them, and wrestle with their legs. I have a cousin my age and he always knocked us all over. He would lie down and we'd try tricks on him—we'd embrace him and he'd lie down, he'd throw himself down and interlace his feet and turn our feet over and we would fall as he stood up. He had great agility in that.

After finishing our work in the field, we used to race on horses. I remember when [I was] around ten, I fell from a mare and wrecked myself all over on my side and my back. And since I couldn't go to the doctor, my mother healed me with *torolico* [sourgrass] leaves. In poverty, imagine, you either got better with home remedies or you died.

In my childhood we went without shoes, although sometimes we used sandals, which were made out of rubber and cloth. The clothes we bought were clothes called mabinga weave, which was the common name given to it, because it was the cheapest little cloth there was. My mother and my grandmother used to make *enaguas* [underwear] for the women with that cloth.[40] For the enaguas they sewed a little strip onto the cloth and that's how they were made for the women. I would

38 Jagüey: *Ficus crassinervia*, family Moraceae; other common name: Jagüey macho. In Camagüey, the leaf of the jagüey is used for liver illnesses. In Cienfuegos, the látex of jagüey is used in casts when broken or for dislocated bones. The leaf is used in a bath for aleviating skin lessions. In Juan Tomas Roig, *Plantas medicinales aromáticas o venenosas de Cuba* (La Habana: Editorial Científico–Técnica, 1988): 53+.

39 Fulgencio Batista y Zaldívar, Cuban dictator toppled by the Cuban Revolution in January 1959.

40 Coa: Taino word, a digging or planting staff, hard wood burned at the point for hardening.

say *enagua* is an Indian word and only we Indios from the mountain used them.[41]

In my childhood the authorities in our community were and still are our elders and we respected the old folks perfectly well, because they taught us how to do the good, because those old folks didn't argue among themselves. They had their system.

Their agriculture was really ancient. They had the *roza*, which is the slash and burn of the land.[42] There was no meadow and the soil wasn't plowed back then. Sowing was done by pick, corn was planted with the *coa*,[43] and each and every one had their own roza. Yuca[44] was planted a lot, and corn, mainly to make the *ducunú*.[45]

My father and my grandfather taught me their sowing system, how to clean the field, how to sow, on which moon and during what season—also how to circle in the forest to plant coffee, all that, although planting coffee, the way it used to be done back then, was also harmful because many trees were slashed and burned. They had to do it because necessity forced my father to work in that. The same necessity made us work since we were little children, from very young on.

We had our own fields. We children worked hard. More than anything [we worked] for our own consumption because nothing was bought. The governments of the past didn't take care of anything, not like nowadays when they buy from farmers. The market was far away. So everything you harvested you either ate, dried or lost. We didn't sell it because everyone planted here and, although poor, we could eat. That's why we had so many homesteads and the Indian people did not leave the mountain easily.

Until the triumph of the Revolution I worked very hard because there was no way to make a living. We planted, but all the produce ripened right there. We couldn't sell bananas to a market. Since my

41 Hernández Aquino, *Diccionario de Voces Indígenas de Puerto Rico* (Madrid: Artes Gráficas Encina, 1977) identifies *enagua* as Antillean Indigenous voice. Today it still refers to a woman's underskirt.

42 Roza: a system of forest clearing and burning characteristic of tropical agriculture and employed by Arawak including Taínos and other Indigenous peoples.

43 Coa: a Taíno word referring to a digging or planting staff of hard wood burned at the point for hardening.

44 Yuca: a Taíno word for *Manihot esculenta Crantz* (family Euforbiaceae). Yuca is a major Taíno crop, around which formed a good core of the Taíno cultural construct. Other common names: bitter cassava (British Antilles); guacamote, huacamote, yuca bitter, yuca hard (Mexico); caxcamote (Guatemala); quescamote (Honduras); mandioc, manihot (French Antilles). Roig, Loc cit (6), pg. 979-980.

45 Ducunú: Montuno name for the corn tamal. Panchito claims the tamal and the word *tucunú* as something proper to Indians.

youth from fifteen to thirty-plus years, I had a hard time supporting my children and paying for their necessities.

I was a lumber worker when I was young. I earned money cutting railroad ties from trees by ax. I'd go deep into the mountains to a place called Toíta to work and fell trees to support my children. Of course, they bought each tie for five *reales* (half a peso). A tie had to be cut on four sides. For you to earn one peso, you had to cut two, and it wasn't easy to spend day after day cutting big trees like those ones. Sometimes I cut up to three, four, and five. But it was so far away that I had to go all the way there every day and then come back all the way to La Ranchería. It was a very hard life and without protection against tragedy.

Now with today's changes, because there is tremendous development, people have left the mountain a bit empty. I'm not saying so very empty, though, because we are in the mountain and we love it very much. My hope is that little by little my families will return.

Since my childhood I knew to take care of it; I knew to love the mountain. I dislike it when a bird is killed, I dislike someone striking an ox because I have never beaten a grandchild, not even my own child. That's something that comes to me from inside and also from my grandfather Cancio, who fought in the War of 1895.[46] He used to tell me that a blow didn't teach anything, that only criminals use it, because you hit your finger and it hurts, and hitting a human being hurts. So since I was little I used to hear that and, like I just said, I struggled with the guys so they wouldn't hit each other.

Our grandparents, my parents, all our people from many past generations were from there, from La Caridad, or what used to be the outskirts of La Caridad, namely Vega del Cedro, La Escondida, San José, Negro Toro, Ocaral, Purialito, La Ranchería—all those were families and neighborhoods, hamlets, bateys, as I said, Indian hamlets. Imagine, my father would butcher a pig, or my grandfather there in La Ranchería, and all families were gifted from it; so when there was an Indio down in Ocaral, he'd send something up from there every now and then, and from here we'd send meat to take down there. Imagine what those times were like, that feeling of community we had, where each one wanted to do more for the other.

---

46 Taíno word, meaning home, also mountain. The bohio, made from the palm tree, is a very extended architecture in Cuba's eastern region

And they'd send boys like me on horseback. When they butchered a big pig, they'd say, "Take so many pieces to such and such people, five pounds, six, ten pounds." Everybody ate. And if someone else butchered, the same happened. Suddenly a boy would arrive at your house with a piece of meat.

Look, this was also done with corn or beans or any harvest, but it has almost been lost today. I ask people, "And why don't we do what we used to do?" Because you see, that way everyone eats, everyone lives a happier life.

Cacique Ladislao, War Veteran

The cacique of our people during my whole youth was Ladislao, my great-uncle, and a very curious man. Among other things he did, he would buy lands, prepare them, and then sell or even gift them to new couples.

Once my father, my older brother, and I, as a young boy, went with Ladislao to clean up a farm on a mountain place called Jiguani, not the one by Bayamo, but one near Baracoa. We were there three months, clearing trees and planting *malanga* [local tuber]. Four of us worked that land, clearing the hill. But it was not malanga for us. Later, it was a different family who worked that land for a long time.

Ladislao rode a small mule. Sometimes he would send me to the store on his little mule. He passed his time talking with me as we worked opening the forest. He was cacique and also my [great-] uncle, and he talked to me a lot. He would tell me, "Because of your intelligence you should follow me, who am now cacique of all those places, la Escondida, La Ranchería." I would pledge to him, "Alright uncle, I will carry on your tradition as cacique of the mountain."

The cacique, I learned from him, is the one who must know what goes on in all the surroundings. The water, the best ways to work the land, the kinds of animals best suited for the place. He would teach me how to raise a cow, the mules, the horses, everything he knew what to do.

He was a pensioner, the last of the old Indians left because he was born during the time of the *mambises* [rebels]. He and his father were both mambises from the War of Independence. He went to the bush with them. He passed much knowledge to me, just as my grandfather did.

I would ask him, "Uncle, how was it during the War of Independence?" He would say, "Look, the Indians were treated like dogs. The Spanish came and would beat us. The Spanish were savage with the Indians, would kill us. If you were not with them, they punished you. If some Indians fought on their side, it was forced upon them, to be at their sword, to be at their feet. They would even make the Indians wash their feet."

On one occasion, my grandfather killed a mule of the Spanish, who were then in retreat. The mule carried a load of *tercerola* guns.[47] They also took the mule for food. At that time the Indians were almost starving. Of the mule nothing was left, not even the tripe, the Indians were so hungry. But they took those weapons and organized themselves in favor of the mambises' war. They joined the mambises with arms. That's what Ladislao told me.

Ladislao was a stand-up man whom everyone respected. He was chief of all around us, of everything, because he had a farm that he split among his children—this parcel for you, that parcel for that one, to both the females and the males he gave them their rights. The way I do is the way the grandfather did; it's what the Indians did.

The Indians, when they ate from a hunt, or in butchering a pig, that food was for everyone. In a house with food, they sent food around to the children and grandparents, over here and over there. That's what Ladislao did, because he was cacique and surrounded by his children, all of whom were intelligent like him.

Eusebia, healer, Ladislao's woman

Ladislao's woman was called Eusebia, an intelligent Indian woman, midwife, *sobadora* [spiritual masseuse], good spiritist. She could heal and she had a son who was a diviner. That one would see a pregnant woman and could tell her: "You will birth a girl." And for sure, he could see the future of people.

We would always go to her house. That great-aunt would purify us before we left. She was strong.

47 *Tercerola* is soldier slang for small version of a Spanish war rifle, the 1895 Mauser.

She would say, "Implore nature to help you, to the sun, the earth, the moon, the star."

That I learned from the grandparents. From them and from all those old Indians I learned a lot. I think that's why I am still alive, for believing in those things normal to Indians.

Ladislao Cacique was very respected. But his woman, Eusebia, was the one who mostly would tell him what to do, since she spoke with the spirits. She would say, "Well, Ladislao, this has to be done, this planting done thus...." Because we have much system that was taught by our father, much system for the planting, with the moon, with the calendar.

There is a lot like that. That's *cábula*.

What we call "cábula"

We call it "cábula" when we communicate with nature. I do it when I plant beans by the night. My father and all the old people, they would study the sky. If there were many stars, they would say, well, tomorrow we plant beans and corn—look how the stars thicken.

They saw abundance in that; that's how they judged things. That system is on its end-times and I wish it were not so. By me it will not end, as I still carry on and so do my children. But there are many who do not. There are now technicians, engineers who say, "one can plant at any time."

Cábula is many things. Yesterday was Wednesday, eight days since we planted beans. We see that there is no rain. But one can hear something. The frog is singing. That is water, we will have water within days. That's a cábula. We say, "I am cabulated [*me cabulé*]."

Life is water. I went to the fields and all the way I am looking for water. I spotted a *bibijagua* [leaf-cutter ant] and then a long row of them, carrying, carrying large pieces of leaf, and I said, "Water is almost here."

The bibijagua is larger than an ant, and she causes damage. That one will eat the yuca, eat the bean, eat the corn, eat the tomato, the pepper. The *ñames* (yams) it will peel completely, the flowers, too...it causes great damage. But, well, that was dictated by nature.

And the bibijagua carrying the big leaf, that is nature dictating that water comes soon. And that is cábula.

Cábula is all communication with nature. Some ask me, Panchito, will it rain today? I laugh to myself, don't you know to look for the bibijagua? Don't you know to listen for the singing frog?

This is a message for the human being. You can know that water is coming and it is a good moment to plant.

Even the stars carry a message. I spot a flying star and feel the cábula. I have to wonder, what is the message? That tells me I will have light ahead of me in my travel. It is a good moment to travel. The Indians understand this thing. But many people, you tell them this and they say, he is a clown. Too many ignore these things of ours.

With that star that shoots through, I can tell a cábula for myself, and I say, tomorrow I can walk to that place. It is a permission to walk and I take with me a clear light.

I said as much earlier, and then this heavy rain is falling here in Santiago, cleansing the bad currents.

My father would tell me: "Learn to cabulate, that is a message that nature is sending you." That's where cábula is. When you hear the birds singing especially strong, I say, "You guys hear those birds?" I say, "I am cabulated today. I am full of signals, because you can hear the birds when they are so happy to sing, yelling, 'water soon!'"

I have a daughter-in-law; she looks at me and laughs and tells my daughter Almeida. "Hey, look what intelligence this old man has!" My daughter Almeida tells her, "That's true, what my old man says requires attention. When he lands on a thing, it comes to be so. It's not whimsical, it's just something he has."

Inside everything, nature

All things cross (blend); even in the Catholic, nature is present. The sun, principally in the holy week, on Good Friday, at ten in the morning, we would kneel before it to ask that the holy sun would spread all his solar rays in favor of humanity. We would kneel for even half an hour asking from the sun and the Mother Earth. At night, with the moon, we did the same, we communicated with her.

Lina and Ladislao's old woman would stroke their faces with their hands and look up to the moon and say, "There is the moon; there is our mother. Look and you will see that it is the face of a woman. She is just like us, [which is] why we have that change of blood that nature gave us. Just like Earth varies her humidity, that's the way we women are."

They used the dew: the humidity of the moon is the Mother of Waters

They would use the dew. In the early morning, they [the elder women] would go to the yuca field to wash their faces. On the leaves of yuca plants, they took beads of morning water—that was healthy. They would drink that water. That humidity and its dew they called the Mother of Water. Into the gardens that they planted they would go and say, "See how the plants grow with the dew." The dew is like that.

That same water they would use to purify a child, the water of the dew, from yuca leaves and from malanga; the water was put in containers, and it would be taken in teaspoons. Myself, when I arrive at a malanga field on an early morning, I drink the water of those leaves. The malanga leaf has a compartment at its base that holds a small spoonful. I take that leaf and drink it and wash my eyes and face in it; I bless myself with that water. That was given by the mother moon, the father sun, the sister stars, because they have names and in their names still we know their gender. The star is female, the moon is female, the brightness -- the sun is male, wind is male, the Earth is the mother of us all, of the whole world.

The water is the mother, my water. And the Mother of the Waters is the moon. The moon commands the water and sends the dew.

The tatagua (witch moth)

A *tatagua* [witch moth] is black and when she comes to the door flapping wings and even alights outside the door, I say, tomorrow a visit comes.[48] I say, prepare for a serious visit, from a chief, or a major friend, a son. . . . The tatagua comes in the day, black, and stays around the patio without leaving. You can say with confidence, I have a visit. Sometimes a big tatagua can come and enter your house. It's a good signal. A bird in your house at night is likewise a good signal; it is a cábula.

48 La tatagua has a popular legend in Cuba. See "Leyenda Cubana 'La Tatagua,'" last modified November 5, 2009, http://leyendas-universales.blogspot.com/2009/11/leyenda-cubana-la-tatagua.html.

The zunzún (hummingbird) gives me cábula

A signal comes with the *zunzún* [hummingbird] that plays with you. I sit there on the porch and the hummingbird comes and circles me. If a plant offers a pretty flower, he hovers to eat. After, he grasps on a thin wire and sits so everyone can see him. He is not to be touched, that bird. Day by day by day, he sits there. The people who go by often see me looking over this zunzún.

Some say, that Panchito has cábula with that zunzún. Some laugh at that, but I say, you believe in yours and I believe in mine. So be it with you that don't believe. But there are also those, many, who say, gentlemen, see, that Panchito doesn't miss. Panchito when he says something about the planting, you have to pay him mind. There is not a tree or plant that man doesn't work. There is no plant he sows that denies him a crop.

Zunzún announces a visit

When the zunzún comes and sits there and gives me a *shshshsh!*, and quickly goes, I say, tomorrow there is a visit. Then, when he surprises or startles me, my body catches his current. Oh, I say, some poor infirm will arrive here. I will await, maybe a woman with a sick child. I tell my wife, something is going to happen. The zunzún presented himself, whistled at me, and left. Something is coming.

The zunzún, when he is sitting, resting, you see him in a light tremor. You look at a sitting one and see his little head trembling, like restless, and I say, I am the same as the zunzún, nervous like that, so I ask him, "What's happening, zunzún?"

We call him "zúmbete," that zúmbete, the people call him "Panchito's zúmbete" and say, there is he is in Panchito's porch. I protect him fully. I look much after all of my nature.

The rainbow, its cábula is water

The old people would see the rainbow and say, there's the water, prepare, see the arch, there will be rain; it is already drinking. My father would say, see the rainbow, he is already drinking water, but he will drop it. In about twelve days it will be raining.

The rainbow brings water and brings cold, the cold of March. In March we saw the rainbow daily. Oh, see the rainbow, I would say.

Today it won't clear up. The rain continues; today its water. Thus they are two, water and cold. All of that is cábula for me, all of it.

From the rainbow you can request his beauty. I see him and say, look how beautiful, and he comes charged with water. So I ask that he does not come so cold. My friend, I ask them, don't come so cold.

I communicate with them because my grandmother Lina, the Mamalina, would tell me, "The best thing about my grandson is that he is in the true reality, that he communicates with nature and with the seven natural potencies."

She would laugh and tell me, "Protect yourself, my grandson, protect yourself, ehh, because you are natural, natural. Whoever goes with you will not fail; you animate them, for you don't fight with anyone, you protect the birds, you guard the forests, careful as you cut even one tree, ehh, even one vine."[49]

Hunting and fishing

In my childhood we hunted the *jutía* [large rodent], and also the wild boar, which hardly exists anymore, but was hunted when there used to be. As a boy I did get to hunt jutías, but not the wild boars. By then the mountain was already being cleared and the wild boars lived in the forest. There weren't many.

The jutía gives itself away by its urine, which is sort of reddish. To get a jutía you climb up a tree and have the dogs below. You knock the jutía, it falls off, and the dog below catches the jutía. We used to eat it.

---

49 There is an interesting correspondence between Panchito's oration and the early episodic poem by Friar Alonso Gregorio de Escobedo. Escobedo describes "indios" of Baracoa in his poem, *La Florida*, 1598–1599, Canto XVI, stanzas 5, 6, and 7, (commented by Jose A. Garcia Molina, in Indígenas y criollos en los primeros versos escritos sobre Cuba (fray Alonso Gregorio de Escobedo y su poema "La Florida"). file:///Users/josebarreiro/Downloads/487-1961-1-PB%20(1).pdf). Wrote the friar, admiringly but also in disdain: "*The rudeness of the Indian is so great that they worshipped the hot ray of the sun, simply for this one being the grandest planet of the West / Of the moon they adored her beauty as she rose in the East and when she peered into their porches, they adored her with pure heart / And then the arch of rainbow shown by God in the heights, infallible word to not forget his humble creatures, they adored with peaceful sense. / Of the great light, of course, the beauty, and of the thunder and his terrible noise, and for the ones they call the Pleíades, they knelt in adoration / They adored the sea, the sky, and earth, and mounds of loose sand and with all that they made war in their spirits to keep the (Christian) God from their own hearts / Their adoration encompasses the streams and the hills and the high mountains, all the fountains—all of that is in the adoration of insipid people.*"

It's been a long time since it was eaten, but it was common to eat it in the olden days, in the days of my youth.

One time as a youngster, we went to a place far away, right next to Baracoa. I went with one of my great-uncles, Ladislao, and my father and one of my brothers, and some of those wild boars passed by. We followed them for a long time, I remember, deep in the mountain, but we didn't hunt any because they went too far into the woods and we weren't prepared for hunting.

When the forests were abundant, there were many wild boars, but the mountain was starting to be cleared and the wild boars began moving away. Nowadays there are more forests again and those animals will be coming back. The woods are being well cared for—the birds and the timber and the beneficial plants. All of that is propagated and is well husbanded. It is respected.

In my childhood we also went to the river to catch fish: *joturo*, *lajao*, *anguila*, *bijaca*, all that. Bijaca is a type of tiny shrimp like a small crab. It's very tasty. But it's been years since we caught any. That was when we were young, when on Sundays we'd go down to the rivers for recreation, to search for things to eat. But after I started my family, when I had my obligations, no more. From then on I dedicated myself to the daily work of my home and my culture, my crops, my animals.

Adolescence and marriage

I was, from a very young age, in my area, I won't hide it from you, much pursued by love. I believe it's because of my sincerity. I never deceived a young woman and so they all invited me to their parties. If I arrived at a party and I had a peso—because acquiring a peso back in that era wasn't easy, but we could make do and we had what was needed—there would be sweets at a dance, and sale of many things. I'd take and pay and tell them, "Go on, eat some sweets here." Everybody ate sweets, candies only cost a nickel, and everyone called for me. I liked to play *maraca*, *güiro*, *tumba*, *marimba*. At parties I was very much the favorite. I liked to sing and I sang well. What we sang were rumbas and women liked to hear me sing. That's how we young people gathered. "Panchito, come sing a little," they'd say, they knew me, some hugged me . . . anyway, that's how I grew up from youth into maturity.

I liked to dress well, wear nice shoes, ride my good horse. The thing is that since my youth I was respected and well taken care of by people and youngsters in my neighborhood. There was no one who hated me, no; imagine how much I was loved in the area that up to now and at my age everyone calls me and I still play baseball with the young kids.

I married at age sixteen. My eldest daughter is forty-seven years old. I'm now seventy-one. After her, I went for some time without having children. That changed. Now, I feel really happy with my children, with my family, with everything, with my father, my father-in-law. I soon made my bohío out of guano and I tied it with *bejuco* [reeds], because there were no nails.[50]

It's still done that way. It's a marriage custom to make a bohío in the mountain for the couple. The problem is that there are almost no palm groves left around there. They've been dying out. Other trees have been planted but there are almost no palm trees. We must plant them. Although not long ago we built one of those round *caneys* [large hut], and everyone who goes there says it's most beautiful. Everybody goes there to dance. The floor is not made of cement but of natural, natural soil. That's exactly how I built my little bohío and in a bohío I feel happy. I'm poor, as I said before, but a poor man with honor and who likes to show respect. I like to be a humanitarian with everyone.

I have ten children. All ten children love me and respect me, and I also have grandchildren. I believe I have around sixteen grandchildren and I'm an old grandfather already. My daughters married a bit early and had children; now I have three sons and a young fifteen-year-old boy.

My children are spread out in several places. I have a daughter living in Matanzas. She wanted me to go there, but oh well! The economic situation isn't easy, it's difficult for me to go there. From our mountain of La Caridad, I tell you that over in Matanzas around Jagüey there are more than twenty Indios. They are working; they went there to work in agriculture. Also in Camagüey there are caseríos of our people.

In Cuba, there are Indios in other parts. I know the ones from Baracoa; they are somewhat different from us, darker, and they live in town. There are others over in the Sierra Maestra, the people from Guamá area, also Indian people from the Turquino and Maisi.

---

50 Bohío: Taíno word meaning "home" or "mountain." The bohío, made from the palm tree, is a very prevalent architectural style in Cuba's eastern region.

Family and intermarriage

Back in those days there was no court here, so people married by their community, before the elders or their parents. They married with absolute respect, for sure, a lot of respect. For everything there was a lot of respect. Even for taking a young group to a dance, an adult had to go along, although we danced and danced a great deal because dances used to last up to three days, eating meat, and feasting, and no one fought. We were truly a community.

In my youth I often would accompany a sister of mine, whom we called Pelaje. All my sisters used to grab me to accompany them out to dance, to go play ball, because, well, I liked diversions. And I'd tell them, "Let's go," and my mother, who was a little cranky, would say, "Where are you going?" and I'd answer, "Mami, let me take the girls to a ball game or to a dance today." And my father, whose name was Juan, would say, "I'm going to check out that place if you want them to go." I would defend them, "Old man, come on, the girls are going with me today."

My sisters enjoyed going out with me. I took care of them and accompanied them. Five of my siblings have died already and I have one left who is sick. We used to dance a lot and go to La Escondida, to La Caridad, to Negro Toro, to San José. All of that was Indian.

That whole area belonged to Indios only. Because, as I was saying, La Ranchería itself had twenty-five houses. But there was more. La Escondida had another, larger amount; San José was next to La Ranchería; and La Lora—all were and are Indio hamlets. Mainly there you find the surnames Ramírez and Rojas, those branches, but there are also others. The Castillo and Paja and the Ara mixed with the Indio and those were the first ones to mix. Some were whitish but they were no longer Spaniards, but rather Cubans with Spaniards, mestizos. Others were darker from the African, and also Cuban.

My grandfather would talk to me. He'd say, "*Carajo* [equivalent to "screw it"], the Castillo came to ruin the Indian race," and he did feel distrustful because he didn't like that they [Spanish] would accost the Indian women. An Indio yes, Indio with Indio, well, we married. Even second cousin with second cousin married in order not to alter the race and continue on. That was the elders' ancient instruction, to try to maintain our people of race. Today that attitude almost doesn't exist anymore. But, up to my generation, yes, the grandparents kept it. I keep it, for I'm married to an Indian woman, but many here hardly do so anymore.

My grandfather used to say, "The Spaniard came to screw up the Indian race." And later the African came, because the Spaniard brought the African, so black mixed with Indio and Indio with the Spanish race. It might be unfortunate or simply for good taste, but that's what happened. Afterward, as things kept developing, we get to what we have today. But since here in Cuba we are all worth the same, we talk about all our origins. Because Cubans come from the three races and we're all Cuban.

Of course, it is valuable to maintain our own culture and that's why I, from childhood on, carry that culture, and as I would play music with my grandfather. Well, just like that I'm teaching my little grandchildren, who already know how to play, already know how to dance. My music is the Indian music, my beat is the Indio beat of ancient times, the one my grandfather taught me, he who told me, "Keep singing together, son, keep dancing and keep planting."

The good sentiments, good human relations—that's my way. And my ancestors were the same, especially my grandfather Juan Ramírez Rojas, the one we called Cancio. My grandfather—I lived at his feet—was the one on my father's side. The other one, on my mother's side, who was called Longo, lived very far away in Bucaral a fair distance from La Ranchería. Therefore I had a relationship with the grandfather on my father's side. He was the one who knew how to truly teach me, he and my father. Also Emilio, my older brother, who helped raise me and is well respected by everyone around here. Opublio, my uncle, is also always very interested in what happens to me. Those are my elders.[51]

I carry the system of those elders, the Indian system, the communitarian system. [It is] communitarian, because I love my little village. I love it and it's not a developed community but a rural one. I call it "community" because we're all close and no one takes from anyone else; we all respect one another there, everyone. There's no push and shove between neighbors and when a child quarrels with another, we know how to give advice; we know how to show respect and

51 Juan Ramírez Rojas (Cancio) is found in the registry of the Hatuey Infantry Regiment, in Roloff.

marriages are respected. My culture is about sharing things without expecting anything back, without self-interest.

Indios and the war of 1895

My grandfather fought in the war. He was one of those old people called "pensioneers." Back then it wasn't like today when many people get retirement benefits. In my childhood, the ones called that were the veterans of the War of 1895. He also wore one of those veterans' hats, a mambí hat. I used to ask my grandfather, "Grandfather, and the Indios fought for Cuba?" and he'd say, "Certainly! Why wouldn't they fight? They fought."

My great-uncle Ladislao and my grandfather Cancio went to war. Those grandfathers of mine fought under the orders of a chief they called Rojita. He was one of our Indios from here, last name Rojas, who joined the mambises in the War of '95. And he was the one who led the group of warriors that left from here to fight against the Spaniards.

Many men from here, from the area of La Escondida, fought in that war. Among them were: Jacinto, Ladislao, my grandfather Cancio, and Serafín. Serafín Rojas was another Indio who was captain and he did indeed take to the war.

My grandfather used to tell that some fought under Maceo and others were chased away from La Lora, and fled from La Caridad to the farmsteads, where they set up camps. He told me that those went onto San José, from San José they jumped over to La Redonda, and from La Redonda to San Andrés, and from San Andrés they jumped to Monte Verde, and later on they went down. They were being chased and went deep into the hills and arrived all the way in Mayarí. My grandfather said he missed the war from there on, because later Maceo was already further down and he couldn't catch up with the main troop. So they had their adventures.

My grandfather told us there was an Indian regiment and they fought even with slings and he said, "Look, we Indios fought hard in the Independence War." My grandfather told me about a little place where they fought, a place called Sao del Indio, and that the Spaniards were loaded with military equipment.[52] I don't know much about that, but I heard Sao del Indio mentioned, and that the battle was fought on the hill over there and that's where the war took place. My grandfather told me that during that combat they spent so many days without food that they were about to fall over.[53]

Also during the war, an Indio sold out and it was he who killed Flor Crombet, a general who came with Maceo. It seems that an Indio dared against his own country, but it's not because the Indios were against Cuba; however that Indio sold out, and he was killed by another Indio.[54]

It's a pity I was not able to collect my grandfathers' memories. They did indeed know about the war. I know little in comparison. I listened to their stories but I retain little about those wars. With them I did go into the hills, into the mountain. I learned what I had to learn about the mountain and I found out that everything has its way and everything has its principle. Their principle was the love for the mountain here, for all that lives here.

In the time of Batista

In the mountain, the Rural Guard was the tough one. One time Batista's guard hanged a man here, right in front of my house. Batista's guards dragged him around and they hanged him in front of my house. He wasn't a man from here—he was a black man and they dragged him all tied up. They committed crimes on him and we saw that. I was about twelve when it happened. A sister-in-law, who was the first to see the deed, started screaming, "Oh! Come and see what they did to this man," and then the whole neighborhood gathered together and

52 Abelardo Padrón Valdés, *El General José, Apuntes Biográficos* (La Habana: Editorial de Arte y Literatura, Instituto Cubano del Libro, 1973): 217–218. Padrón describes the participation of the "Yateras aborigines" under command of Antonio Maceo in the Battle of Sao del Indio, August 31, 1895. The Rojas-Ramírez Indios who had joined the Hatuey Regiment distinguished themselves in this battle. Padrón Valdés writes: "From the right flank came the attack of the Hatuey Regiment, then commanded by the Dominican general, Dionisio Gil (Noni) and which was composed largely by the ancient Yateras guerrillas, who with arms had crossed to our forces . . . These guerrillas behaved heroically in their debut as patriots, liquidating the (Spanish) artillary troop and taking their cannons. . . . ." (Translation by author).

53 José Sánchez Guerra, "La Capitana del Regimiento Hatuey" (Guantánamo: Revista El Mar y La Montaña, October 1998). In this fascinating article, the Guantánamo city historian gathers the genesis story of the Hatuey Regiment in May 1895 after a ceremony by the midwife and seer, Cristina Pérez and Pérez, in which the spirits of ancient Taíno caciques appear.

54 Juan J. E. Casasús, *La Invasión: Estudio Crítico-Militar* (La Habana: Academia de la Historia y Academia Militar de la República, 1950): 48–49. "In the heights of Alto Palmarito, in an encounter with the Yateras Indians . . . dies . . . Flor Crombet. He is killed by the Indian Rojas, 17 years old, who a few days later is incorporated into the Liberation Army." It is possible that this oral history confuses two versions of the same incident.

went over. It hurt us and we held a wake for him, although the guard didn't want us to. But, what were we to do? We held the wake because they had left him like a dog on the royal highway. They just left him thrown there and we held a wake until his relatives, who were from the town of Manuel Tames, had been notified and picked him up. We kept vigil over that man as he was a human being, and so the vultures would not eat him. I do believe that's something all of humanity is obliged to do. Later, they wanted to imprison one of my brothers, accusing him of the crime.[55]

Those were bad times. I myself was pushed and battered by Batista's guard. On account of all that, we were rebellious against Batista, until the Revolution came. Batista's time brought us lots of suffering. We were on our own and things were very tough. No one studied, hardly anyone could read, although we did know what was happening.

Another time I remember that the guard of the dictatorship caught an Englishman (Jamaican) in the mountain and they made him flee from there. He crossed the river of La Escondida and they fell over him from behind and gave him several blows. Those were crimes they used to commit in that past government. When I was a child, well, we were fearful, because of those people killed and hanged, and we were afraid to leave from our homesteads to where the Rural Guard might see us.

I was a little boy still, perhaps eight years old, around that. My father had a pig, called Shotgun, fattening. One day a mounted patrol stopped by and they wanted my father to butcher the pig and fry up some meat. My father said no, that he was still fattening the pig, but they pushed him so much that he had to kill the pig to make them food. My father also had a cowhide set aside to make a bed, and they asked him where he had stolen that cow. My father replied that he was no thief, that he had bought that cow from a man called Pino Arias, who sold him a heifer for that purpose. The guards kept insisting, but he said no. The matter rested, but always with danger.

On one occasion I myself, as a child, was pushed around by the guards. It was at the schoolhouse we had, which wasn't even a school because no teachers came. The guards were offering lunch to buy votes

---

55 Rural Guard (Guardia Rural): a mounted, rural police apparatus organized in Cuba during the American occupation from 1998 to 1902. It was characterized by a highly repressive practice, including theft, torture and assasination, over the civilian population. It operated until the advent of the Revolution in 1959.

for the election. We passed by and my father told me not to eat because we were not going to vote. I left and went to sit under a mango tree. The guards insisted I eat, but since the elders had told me not to I said no, and they started pushing me and lifting their rifles. Anyhow, little by little, we distanced ourselves and continued on our way. Those were things from my childhood during the time of that government.

Struggle in the Sierra Maestra

As early as 1957 we heard that Fidel was in the Sierra Maestra. We didn't have radio but we farmers knew what was going on. One of my brothers told me there were rebels in a place they call Las Trincheras, and that Batista's airplanes were bombing the area. That hurt me. I told my brother, "Look, let's block the way so the guards can't pass." In order to prevent the guards from passing to where the rebels were, which is where Raúl [Castro] was, a place they call Bejuquera, we cut a very large tree so that they would have difficulty crossing and wouldn't get here. We also placed a very big rock in the way. But the truth is the guards never came up that way.

One of my brothers and I would ride on horseback and take the horses loaded with food to the camps. I was around twenty-two, young, and new. We'd take food to the camps for the rebels, and we ran errands. The *comandantes* [major rank in Rebel Army] were Peña and Fajardo, and [there was] also a lieutenant Labrada. Peña and Fajardo have already died, I believe, but Labrada is alive. So, because of those activities, my children tell me I was in the "struggle." Well, I always have fought for my country, in the sense of trying to help it with what I can. It's the sense I have and I have it by culture. I'm not saying to you I was in combat, because all this took place on a most protected mountain and the [Batista army] guards were afraid to climb. It is one of those hard trenches, and by then they wouldn't cross over there. But yes, we were there supplying food, even though the guards had already stopped coming through; there was no way they would have come, because we were well backed up.

We did this out of love for the rebels, because we already knew Batista's people, who were the same old thing, and who for us were what the Spaniards had been for our ancestors. But Raúl and Fidel—

well, we knew they had something different; they got along with the guajiros of the mountains.

I love my country and I would rather give my life before turning against it, because the state has provided us authority to continue practicing this Indian culture I have, so beautiful and decent. Because culture is not only the music—culture is also planting, also the human feeling within that entire philosophy. That is my culture; that is what I am.

Guajiros in Havana

In the early time of the Revolution I went to Havana. I mean, not only I went, but half a million of us *campesinos* [peasants]. Fidel made a call to the farmers of the Sierra, and a few of us from here went: Emiterio, Pedrito, Gaspar, Severo—all dead by now.

A few of us Indios went there and we strolled around with Ché. We went in special trains and were provided with good horses. It was a great caravan from here from Guantánamo to go sign in with the Revolution. Over there we toured with the leaders and were very well cared for and not only by the government, which was just beginning, but by all the people in Havana.

We went first to the University of Havana, but stayed in private homes. People came looking for us. That was very nice. People came looking for us and *bam!* we were family. You took me to your home and my cousin would go to another home and so on. People with people, as if we had known one another forever. And this in Havana, city of one million, where few know each other. And Revolution does that, *cubanía*, our Cuban character does that. People sometimes complain, but the truth is the Revolution really lifted up our *cubanía.*

That time I personally met Ché, and I met Camilo. Well, that was wonderful. Camilo was and is a very popular man, very popular. Camilo was a kind man, with a tremendous heart. There in Havana he hugged us, and imagine, he understood us. For the first time in the entire lifespan of the oldest among us, no one had ever known a leader who got along with us. And among them were campesinos, like us, and many even with the machete hanging from their waists. Well, that is a true story. That was to live an entire lifetime.

When we arrived in Havana they were waiting for us. Raúl talked to us first. And there they told us, "Look, this is Ché, this is Camilo, this is Fidel," and they started getting to know us personally, shaking hands with every one of us, the same way we do in our community. We left with Ché and Camilo on horseback. They said, "People from Oriente over here, up front. Guantánameros, here." We stayed seventeen days. And that was homeland.

How am I not going to love my country? To say "*patria*" [homeland] is to say my siblings, my uncle, my mother, my father, my friends; to say "homeland" is like saying to you that I love you, to my children that I love them. Because all that is homeland to me. That's patria and I believe that if that humanity isn't there, there are no friends; and the true friend is a brother, brother, brother, because we all feel the same and this one knows a little more than that one, and the other knows one thing and another knows something else—anyway, all together we can.

After the triumph of the Revolution

After the triumph of the Revolution of 1959, life around here started changing, changing for the better. Children are now able to study. All neighborhoods have schools; there are clinics; there are doctors. Because the most important thing is health. Here in this country that's the main advantage: we get sick and we go to a hospital, and the assistance is given for free.

Because in the past, your child got sick and it would die on you. Since you didn't have money to pay the doctor, you could die. If you hurt yourself with an ax, you could die from fever. I cut myself with the ax in the old days and I know what it's like. Later on while working, a tree fell on my waist and I was unconscious for seven days at home, because there was no hospital to take me to on the mountain.

But not now. Right now with this little cold I have, the doctor of the 120 families, one who is in La Caridad de los Indios, has already seen me. There they listen to my complaint and the doctor says, "Come, let me prescribe some medicine for you." Of course we have medicine here, medical and natural, for anyone. In that regard I say, that is humanity, that's the most precious we have here in Cuba. Health and education, that's the most beautiful, and we take care of doctors like our own children. We take care of teachers like our own children.

At La Ranchería my brother has been sick for a long time; the doctors have slept over at our place, observing him. Up to three doctors one night: the doctor from El Pinal, the doctor from Caridad, and the doctor from Viva.

Look, this is history and since you are writing my story: I had a daughter in Santiago de Cuba, one who works with me in the traditions, who had cancer and now she is well again thanks to the system and the Revolution's attention on medical assistance for poor people like us. She fell gravely ill and I didn't know what to do, but they assisted her, she had surgery, and they treated her thoroughly and cured her. What would I have done with that daughter? How would I be able to pay for those three months at the oncological hospital in Santiago, and for her being healthy now?

Well, now how can we pay back the state for all of that? By sowing plants, by raising animals so the people can eat. That's how we pay the state for it, because I have to thank the state for all the treatments that girl received, who today is enjoying her life. I'm thankful for that.

We are humble people but we have our human dignity coming to us from long ago, and we know how to be grateful without shame. We are guajiros from the countryside, from the mountain, and we have a lot from our elders. Although it's true that many people from the village still look at the ones from the mountain and, as one example, say as if to belittle us, "Look at those guajiros; look, there come the guajiros." In any case, I want to say that we are not offended at all by that, because we are in fact guajiros and much of what the guajiros do is Indian, too.[56]

Volunteer in the army

I volunteered in the army for about two years, and then came back to my neighborhood again as a farmer. Help was needed and we went to give help. That was in the time of the bandits entering through Las Villas. They mobilized us here. Many, many of us comrades went, just

---

56 See Irving Rouse, "Circum-Caribbean Tribes," *Handbook of North American Indians* 4, Bulletin 143, Julian H. Steward, ed. (Washington, DC: Smithsonian Institution, Bureau of American Ethnology): 530+. "Several sources mention a term, 'guajiro.' This may have applied to the common people. Las Casas however lists the guajiro before the nitaynos." In the Wayuu language (Venezuela), *waxir* or *waxeri* signifies a "man of his home" or "man with house" (extended family). Noeli Pocaterra, Wayuu Tiger clan mother, in personal conversation with author, 1991. Jose Juan Arrom, in personal conversation with author, 1991. José Juan Arrom, in "Aportes Antillanos al español de América," *Revista Areito* 7, no. 27: 12+: "... guajiro ... from the Taíno ... 'our compatriot,' or 'our companion.'"

as we are doing right now, to defend our country; we are defenders wherever people might be entering by force. So we went gladly, of course. They were saying it was a tough battle against people infiltrating by motorboats from the north. So we prepared ourselves for whatever might be, promising to make ceremonies and leaving offerings on the altars. We were ready, but in those days of mobilization, we helped in the countryside above all. At that time, we cleared a lot of land. The army cares like that for the farmer, helping without asking for anything in return and people get along with the army. The Cuban of today is ready to defend. We gathered tremendous experience. What pride one feels when defending the homeland!

Medicine in the Revolution

I want to say this before talking about my own medicine—well, the way we do things on the mountain. This is so no one gets confused. Because, listen up, we must be grateful for the things the Revolution has brought to us—labor and delivery, for example. In former times women delivered at home. The midwife helped, but women had homebirths at substantial risk. There was no medical support, and I tell you truthfully, women suffered. It was sad how women suffered and died. There's my own wife who almost died. We had to carry her down on a stretcher from La Caridad de los Indios to Manuel Tames, where the doctor was. That happened in 1959, when the Revolution was only beginning. The hospitals weren't there yet. If it were like today, no—today in the mountain there is a hospital everywhere. Back then, however, there was nothing. When she started labor, the midwife was with her. But she had a breech birth and she almost died. But not after that. After that the Revolution was happening already and all women gave birth at the hospital. Today women don't go through such labor to give birth because they deliver in the maternity ward and there are maternity hospitals and everything is provided.

But in those moments life was difficult. The Revolution was beginning and there were no doctors. Doctors that were here in Cuba were very, very few and the majority left; but now we have them in the thousands, thousands of doctors—even relatives of mine are doctors today. I have cousins who are doctors, [who are] male and female nurses. From La Caridad there are two doctors and a cousin of mine called Neri Ramírez. I also have two cousin nurses here and another cousin who is a nurse in Matanzas.

I say this because I am all for science. I like the way a scientist knows. But our own way, since it comes from the heart, since it comes from nature, also has its science. It's simple; it's the science of the Indio, and that's what I'd call it. The old guajiro—one could say the old folks—had their own science, too.

. . . for me, those plants are like a village; they are families in the countryside. And that's what is most essential, that respect. Because those who know what that is, know how to ask for permission, know how to make themselves recognizable before nature.

Nature

Respect is essential

The world has powers. Our Cuban earth is powerful. Everything has a spirit; everything has its own energy. That we know. We carry it in our families, that respect, because that is the essential.

In our ways we have the medicinal herbs, and in the work with herbs you must show respect. When I'm going to pick little leaves, I think about that leaf. As I'm going to snap off that branch, I ask her for permission, so she gives me the power needed against any disease, for any cure I might need her for. I approach her with natural fervor, the fervor of humanity, so she gives me the medicine to heal, to me as well as to anyone else who might come seeking. If I take a branch to sanctify anyone, I ask permission from the plant. Also when we do baths, I cleanse the person with different plants.

Respect is most important. We also have a system where you have to buy from her, give something back to nature. As we go to pick a plant, we offer some pennies to the stem, in order to obtain more experience, to obtain the most strength the Earth can give us, to take as much power as the world has in that plant, and in order for her to feel, to know it's true that she is going to heal a sick person.[57]

That's my system, because there are many who don't [do it], there are many who go and rip off a leaf and *zap!* tear it without thinking. And in my view that person has no real respect. That little plant needs to be cared for, because in our ancestors' traditions that plant is almost a human being, since it feels just like human beings do, because if we tear off a piece of her, she withers right away. But if we give her what we have to offer so she may live, she gains strength. Same with water: when you water a plant, she responds. Again, she feels that human communication.

That feeling comes to me from my grandfather [great-uncle] because that's how the elders used to think as well. That's how many in the family I have used to think, all those who are old by now. And now that's what we tell the new generations, to do what we do in order to have faith in the land and in what the land produces for us human beings. That's a system I have of human feeling for the plants and for

57 José Barreiro, "Indians in Cuba," *Cultural Survival Quarterly*, Fall 1989: 51+. Doña Dora Romero Palmero (78), formerly an active midwife and herbalist of El Güirito, and Aleida Hernández, also an herbalist, from Punta Maisí, both of native heritage, mentioned to me the use of "tobacco seeds," in reciprocal function as the "pennies" mentioned by Panchito.

everything there is on Earth. It's very nice, this natural feeling; this give and take is a feeling that comes to us from belonging to the Earth, because we are Indios.

If, for example, right now I need this little plant and I ask nature to lend me, to give me that, I offer tobacco to her, I pay her with a penny. I pay her with anything. It's a form of respect, because that object seems alive to me or as if it were a human being. I repeat that, for me, those plants are like a village; they are families in the countryside. And that's what is most essential, that respect. Because those who know what that is, know how to ask for permission, know how to make themselves recognizable before nature. Well, that's a system that I carry.

Natural medicine

We pick different plants for medicine. There is basil, *salvia*, *romero*, verbena, and *pasote*.[58] Well, there are so many. For instance, when we have a headache, we boil the leaf of *guaramo*. We gather cedar bark and take it for different ailments. The cayenne plant, for example, we take good care of it, because it's what you use when you don't go all the way to the doctor. For fever, flowers and leaves of the sour orange—all that is good for fever. We also use *rompezaragüey*, and well, many more medicines, many that I don't recall now, but anyway, with time, as you practice, you begin remembering medicine. It's because with natural ways you watch things as they are unfolding, not so much to make lists of things. In my school in the mountain, life itself teaches you. As you see things in nature, they relate to each other. That's what it is. They are relatives, they are compatriots of ours because they help us.

The truth is that in the mountain, the most appropriate medicine we have is what the students now call green medicine. We have all that treasure given to us by nature. Let's see, some of them: there is copal, of utmost importance, which is very good for headaches. The bark is for the cold, to drink as tea. And from the resin of the copal you make small balls like tiny pills. When you have a bad cold, you take two or three of those tiny pills, and swallow them— the same for headaches.

Copal is also good for pulling out thorns, any thorny prickle that might get into your foot, any kind of thorn, or if a bug stings you and it won't come out. You stick a little copal patch on it and that pulls the

58 Romero (*Rosmarinus officinalis)*, medicinal plant; salvia *(salvia officinalis)*; Pasote or Apasote *(Chenopodium ambrosioides L)*.;

thorn out without having to cut it out or prick it. For that thorn to come out, you stick on a copal patch, leave it on for a couple of days and when you pull it off, out comes the thorn already stuck to it.

If you have a splitting headache and you take a bit of that copal resin and put it on a guaramo leaf, or on a yellow yam leaf, you stick it on the sole of your foot and that headache goes down.

My grandparents taught me all that, my grandparents who didn't leave the mountain, where I learned to do all those things, and they are real because my mother and my father followed them conscientiously. Copal does all of that, and out of all that most importantly, when they made a ceremony, they used copal for a prayer, for smudging. Copal is essential in our sacred ways.[59]

There is mint—another simple thing, but very beneficial. So when you have a very bad stomachache you drink a shot of mint tea and it's very good for digestion, and food that might otherwise harm you will agree with you.

I even have a brother who has been suffering for a long time from, well, the doctors sometimes say it's cancer, but I say that perhaps they are wrong because he's been sick for so long and if he had cancer he would already have died. We give him mint juice often and although he's not perfectly well, he's been staying stable, because for years and years and years he's been sick and we give him all that medicine, all those little plants I'm telling you about, all that we give him.

The cleansing smoke (smudge)

That feeling of respect is the same you carry when cleansing and connecting with smoke, because the main thing is you have to be genuine. The cleansing smudge is made with copal, with tobacco, with other offerings.

Look, at my house we used to have an apiary when I was young, and my father and my mother and my eldest brother used to work it. And it was perennial that every Friday they'd smoke the entire apiary with copal and they said that copal was something that prevented any type of bad influence from entering the apiary. That way there would be no evil eye.

59 Dora Romero Palmero, resident of El Güirito, in interview with Barreiro and Hartmann (1989), walks to her "brother tree," a thirty-year-old copal in her yard. "Copal is very sacred," she says. The copal resin is used as the principal sacrament in ceremonies associated with the ancient Maya calendar and other Indigenous spiritual traditions in Mesoamerica.

When we see something strange, we say it's a trick joke, or when an animal is sick we call it an evil-eye spell. Or when you find that there is a person who has the evil in the eyes and might be looking to cast that evil. These are things that come from bad people, people who want to do harm to you. Or also, it can come from nature, from something you've done wrong or you have not fulfilled. In those situations my mother would request for that problem to have a cleansing smudge with copal.

Tobacco smoke is used too.[60] When there's a person needing to be blessed, you take a large leaf from the tobacco and burn it to make smoke, and the smoke is an offering you make. Mainly for San Lázaro, if you carry San Lázaro, for example. You smudge him and ask him for permission to take away the bad influence. Moreover, tobacco is an aromatic leaf and that's why people smoke it so much.

I have a sister-in-law and she has a little plant she takes care of very well. They take the seeds, toast them, and make medicine for colds. For a very bad cold you might have, you take a little drop like that from the pan in which you toast the seed, you peel it and oh! it works, it works. Tobacco leaf is also useful for inflammation. You put the tobacco leaf on it, let it wither, and it relieves the inflammation.

Then, too, also as an offering, when you pick a leaf of an herb, you can offer a cigar, or you may offer it to San Lázaro or any saint or spirit you might like that you might adore or you might understand.[61] You place that cigar there as an offering. That's the first offering you make for a spirit, for a plant, or for a saint: a cigar. For nature, the ceremony is of tobacco. The prayer of the Seven Powers is like that,—with tobacco, the greatest sacrament.

The first water or rain of May

The old folks used to say you should drink all that cup could hold of the first water of May and to get whatever you could fetch in a water pipe or bucket in the patio. If enough water came down they collected

60 Tabaco: Taíno word for *Nicotiana tabacum*, but also the name for the rolled tube. Tobacco is a ceremonial sacrament used by most Indigenous nations in the Americas. Its use, both sacred and profane, has extended to Afrocuban religions and society in general.

61 San Lázaro is a Cuban *santo* or saint, emergent from the people and not formally recognized by the Catholic Church. Fariñas , Ibid. p.p.124-127. San Lázaro, a womanizer prone to ire as well as benevolence, is a peculiarly Cuban saint. Fariñas correlates San Lázaro with sustaining in popular imagery the spiritual (*cemi*) essence of Yucahu Bagua Maorocoti, Taino Supreme being, also incorporating the *cemi* named Corocote, similar in his womanizing ways.

it in bottles and those who didn't drink at that moment had to drink when they got back. So they'd arrive and drink the water of May, the first, very first water of May.

They'd say to it, "This is the water that ended the drought, the holy water of the first day of May," and well, that's a system I still interpret and follow. That system doesn't end, which means that drinking this water and doing many things with this water is a way of life—because that's how the Earth gets her fertility.

Water belongs in our ways. We appreciate water. It's a sacred thing, tears from heaven, tears from Earth. All waters must be taken care of. They have their saint; they have their spirit; they have a mother. Water from heaven, waters from Earth, they're blessings that come together within us. Water is in the Seven Powers.

The lightning stone

Another thing we do here is we respect the lightning stone.[62] In my temple I have one and I also carry it in my pocket. Sometimes I go to the field or I leave to go somewhere else, and I look at it and I feel inside of me, I feel my lightning stone giving me strength. When I see the drought severely hitting our crops, I hold the stone and I ask it for help. I wouldn't want lightning to take me away, but that comes from beyond and arose from nature. I have it as an offering, as a dish, and I dedicate it to my temple there, because a lightning stone comes from the atmosphere, and I see that water also comes from the atmosphere. The lightning stone—people don't know the importance that has. We Cubans, if we want rain, should never leave the stone, which calls lightning, calls thunder, brings water.

It's a small black stone like this, longish, found there where lightning strikes; just a little later you go and see that little stone, because it emerges and you pick it up. You take a lightning stone and tie a thread to it—sewing thread works—and you can throw it into the middle of a stove and the thread doesn't burn. I have done that test and you can trust it, being a lightning stone, because with any other stone, I throw it into the fire and the thread burns, but that one doesn't. Tied

62 Belief in the medicinal and energetic powers of the lightning stone, or *piedra de rayo*, is found throughout Cuba, and is a universalized tradition noted in American Indigenous and in other cultures. It is particularly associated with the deity Changó, of the Afrocuban religion "Rule of Ocha." See Fernando Ortiz, "Las piedras del rayo: Folklore religioso del cubano," *Revista Archivos del Folklore Cubano* 1, no. 2: 96+.

to that one the thread does not burn, no matter how strong the flame. That's a lightning stone, which has something that doesn't let anything attached to it burn. The lightning stone is protective.

We use it to bring rain and for other things. When someone feels sick, I rub the stone on the person and the stone transmits strength. There is also affection, you know; if I have it there, I have faith in it. It's also the affection I have for it. For me it's a person, because I hold it like a god. And that attraction of rain is to bring lightning so that it falls on the earth. And it sounds, when a lightning bolt strikes a palm, a tree, you look at the weather, and there comes the rain.

It's said of this stone, by the elders from the past, that this stone emerges every seven years, seeking a lightning bolt, a thunder. It hits, and a little stone enters right there and later comes up to the surface. I have one and I have it on me. Or I leave it on my altar to purify it. Sometimes people come to me to be healed, because I'm sincere and nature comes to me. And I heal with my knowledge of things. I do the ceremony and the work with that stone: I place it there, I pour a little alcohol on it, and I baptize people with that stone and I do my spiritual work. I invoke that stone, yes, to take away what's bad, the disease the afflicted one has, person or animal. That's an old system of ours that I have. I'm not the only one carrying the lightning stone, but not everyone has the understanding and the affection I have for it.

Ceremonies and chants

If I'm going to do work with a sick person, I sing. Songs have their power and the natural powers enjoy them. And I sing a lot to my spirits and my saints. A strong *santo* [saint] for me in this work with the people in the mountain is San Salvador. I have a song for him that goes, "Save him, San Salvador / save him, San Salvador / save him San Salvador / from injustice and betrayal." In that chant I ask San Salvador, this spirit from here that we call San Salvador, to save the sick person and to give her his power. That is San Salvador and he helps me; he comes to me in my work to cure an illness.

Now, if it's a spirit I see holding on to the person, well, then I sing "Forgiveness": "Forgiveness, my Lord / forgiveness my mercy / for they who asks for forgiveness / forgiveness shall be granted / Virgin Mary / who is God's friend / child in arm / asking for forgiveness."

I go back and sing, asking all the mediators, who are things of the world themselves, to lift what is afflicting her. "Forgiveness, my Lord, forgiveness, my mercy, for they who ask forgiveness, forgiveness shall be granted." Because that person may be carrying sorrow or guilt, and the world has cast something on her, from man or from animal or from an element itself, which is bothering her, weakening her. It might be her fault or someone else's, or a curse that got a hold of her. Therefore you have to ask for forgiveness. You have to know what belongs to what.

So, those are the works of mass, of ceremony, and when I'm going to raise a prayer and ask the Elder Lord of the World, I raise it with that, with the song that he who asks for forgiveness—well, if a spirit comes, for example, and asks for forgiveness, forgiveness shall be granted, so that it moves onto its glory. It has to be done or it'll continue bothering the pained being.

There are many spiritual songs I have. I tell you that a lot the songs left to us refer to the names of santos. But the fact is that the connection is made with whatever spirit comes to me in my dreams. "Save him, San Salvador," is one of the songs I sing the most. He is one of my santos. I also sing, "Listen to my song," to the three saints, because when I have to ask for something, there are many ways and I have lots of songs, although at times I'm stuck and they escape me. I also have a song that says, "May San Salvador bring me the health God sends to me." That's for a person who isn't well, and I'd like to sing, so for example I sing to the sick one, "San Salvador may give you the health God sends to you." It means that I'm looking for health; what I'm asking for with the faith I have is health for the one I am healing, in order to heal.

The one conducting the healing has to protect himself, of course, so the bad influence doesn't fall upon him. First, I have my bandanna. I tie my bandanna around me; when I don't tie it across I tie it to my waist, and if not, I put a little lotion on me, and I rub myself with a bit of alcohol. With a fistful of alcohol you're protected from bad influences, from electric shock. We call an influence a "*corrientazo*." That influence is like being struck by lightning. Lightning comes and you're there and the lightning is going to strike you. That's an influence that comes to you; it's an influence and that's the gift I have, my gift. As I say to you, my way is that I believe in what I do.

Now, all medicines reveal themselves to each other and one leads to another. Everything must be respected. It's not always easy to know their relationships. Myself, I ask sincerity from everything I sing to. I sing to the Lord brought by Christians, just as I sing to the spirit of the *ceiba* [kapok tree], of the *guásima* [West Indian elm], I sing to a santo, I sing to the Virgen del Cobre,[63] the one who encompasses all that is ours, as I sing to the plants. But when I'm going to pluck a little one of those remedy plants, I call upon all those elements, since the plant is alive and I'm going to sacrifice her, so that they take possession of that little plant and give to the little plant and request of her to give her power away. If I'm going to sing to a little plant, I say, "Virgen de La Caridad, give power to this plant or help me so that this plant gives me her power to heal." Virgen del Cobre is almost always part of my devotion.

For anyone having a headache, having an earache, if that plant is recommended for it, I encourage it, so it goes with that saint, with that faith. That has validity; it even serves to have mutual understanding. If I have to say, "Rompezaragüey,[64] give me the good fortune to talk to a loved one," to me it appears that this little plant is the one calling one's companion. And that's like calling on the telephone, so this companion talks with me, as a brother, which is what we human beings need from each other. For myself, I have that faith, that devotion.

And there are many times when it has worked for me. Even with animals. Because I've even gone into the pasture of an angered bull and I've said my prayer, and as I've told my children, "Listen, you pray these words." So I did: "Ferocious animal, get away from me." It's something I carry from my grandparents' teachings. They'd say, "Evil, ferocious animal, get away from me. I thank you for leaving." That's how they'd do it with an angered bull.

Now, if I am going to give a feast, if I'm going to have an activity, so no one starts a fight I start by praying, for example. I say my prayers so all those brothers and sisters coming to have a good time with me be equal, that we be fathers, mothers, that we be brothers and sisters, that we all be sons and daughters, and certainly here in the mountain

63 Olga Portuondo Zúñiga, *La Virgen de la Caridad del Cobre: símbolo de cubanía*, Agualarga (2002), Indiana University (2008): 51+. "The image of the Virgin Mary (Virgen del Cobre) is incorporated by the aboriginal . . . as one more Cemí." Translation mine.

64 Rompesaragüey:. *Chromolaena odorata* (L.).

everyone assists me and listens to me, that if they come, here respect is required among us.

Children ask me, many women ask me, "Panchito, how do you feel when you are singing?" When I sing, I feel complete; I feel connected.

It was my grandfather who'd sing to me about our Indian heroes from here—chants of our people, spirits of Hatuey and Guamá,[65] who were rebel caciques that rose up in arms in the mountains of Baracoa. The Spaniards made Hatuey disappear and it was Guamá who remained for several years, but they made him disappear as well. So my grandparents, they used to tell it, therefore notice that those chants go like this: "Yara, Yarisa is leaving, she's leaving with the Cacique Guamá. Yara, the India is leaving / with the Cacique Guamá," because Guamá was also called "cacique."

Our music is Indian, and the beat is the Indio beat from long ago, the one my grandfather taught me. In Cuba there is also the African drum beat, which we also embrace. But our drumbeat from here from the mountain of Caridad has always been different. It's the Indian beat, which is from here; it's the beat of the Taíno dance.

A Havana man came to interview me once, and upon hearing my beat, he said, "I've found the Indian key." And it's true, because that was the music of the *areito*. We'd form a choir to sing. And our dances are also from here from the mountain, just like the *ruedo*, the *carril*, which we'd do right and left. One couple went back and forth, the other one behind turning around. I'm going to demonstrate it to you, the dance of the Mamalina, which is the dance I do, the dance of the Mamalina. That's the dance my grandparents taught me. That's where you can see what is ours, our areito.

I say it's true that we took many things from Spain and also from the *Congolí* [Congolese], well, from Africa. —the *changüí*, for example. Because already during the Spanish war, they danced the changüí, which is a most Cuban thing, with roots Indian, roots Spanish and roots African. But the circle dance, that one is indeed more like the old areito, where we went one behind and the other ahead in a circle. Because every dance in my culture has the same beat—*chaca, chaca,*

65 Hatuey, d. 1511, Taíno cacique from La Española island, retreated to Cuba to organize the native people against the Spanish. Captured and immolated by the Spanish, Hatuey famously refused Christian baptism while warning his people about the conquistadors' thirst for gold. Guamá, another Cuban Taíno cacique, took up arms against the Spanish through 1532.

chaca—with the maraca made of the same bamboo, which today we modify with the *güira* and the *guayo* [gourd instruments]. Our güiro of yesteryear was made of bamboo, with corn or little pebbles inside. That certainly was from long ago, and not so long ago it belonged to the Indio.

Ours is a culture of spirit, spiritual. "Like that," as my grandfather would say, spiritual and joyful. And you see it in those pretty songs, like that chant I sang for you, or like the caringa: "Take, take, take Caringa/ old stick made of jeringa." [66] Those funny songs belonged to the old ones, those very old chants from the mountain, because pay attention, listen: "Caringa, old stick, take it, take it, take the Caringa, old jeringa stick." Those are Indian words.

Jeringa, the jeringa was made from the male jíbaro pig. They made jeringa out of braided pig tripe, and that was the jeringa; then they smoked it. The pig tripes were well washed and they stuffed them into each other, calling it "*andullo* [tobacco plug]." So the jeringa was the same or almost like the andullo, and it seems that's where some of those funny songs of the Indios emerged. "Take it, take it, take the Caringa, not old stick, not jeringa stick," because listen to the words, "Take Caringa, not old stick, jeringa stick," which means "limp stick." There is also a caringa music different from mine that other groups are playing now, but it's not with the beat my grandparents used and gave us.

Here in the mountains while planting or weeding or harvesting, the old folks also liked to sing; it was a custom. They sang the *quiribá*, the *nengón*, the *manajú*. As they were harvesting, working in the field, they sang to each other. At this my grandfather was good, indeed. I'd stand on a hill over some ravines far away, and I'd hear my grandfather singing, "They say that manajú is useful for making remedies / I am going to buy me a nickel / so to enjoy good health / Mama manajú, pylon manajú, manajú piylon, mama manajú, pylon manajú / Over there on that high ground / they killed a big rooster snake / pulling from his insides pop / seven mares and one stallion."[67]

---

66 Caringa: popular dance and song style of early Cuban origin, with some roots in the Canary Islands and largely formed in Afro-Antillian culture.

67 Nengón, quiribá and marajú are noted as Cuban dance music traditions from the regions of Manuel Tames and Baracoa.

Many of those chants were improvised, from times long past they made up all these songs. While planting, you'd hear them in the ravines planting. Or if not, you'd hear the quiribá. "Quiribá by night and day, Quiribá, Quiribá, Quiribá in the morning, wake up Juliana, Quiribá, Quiribá / to pack up my luggage, Quiribá, Quiribá / I'm ready to go traveling, Quiribá."

And that's what those Indios seem to have called out while leaving at dawn. "Grandfather," I said, "and quiribá, what is quiribá?" And my grandfather said, "Quiribá? That was born from an Indio. "The quiribá," he said, "was the one who called out songs and people responded."

Religiosity of the cruza'o

My grandparents told me I was a cruza'o, I was going to be intelligent, because I sang many chants, of saints and of nature, and I sang things even without knowing it. All those chants just came to me and so my grandfather used to say, "This one's going to be religious. He is someone of power or a *santero* [Santería priest]. Listen how he's singing and playing the drum."

So this was born within me. I am nervous, but I used to get more nervous. Even as a child I'd get tremors. So I'd go to a sister-in-law's, who worked as a spiritual healer, and she'd sanctify to purify me. She's a cousin of mine called Deodora, very old by now, and she says, "You have to give adoration to the saints. You have to light candles for them because you are a santero, and you are cruza'o, you are a religious person of ours."

The cruza'o among us has something from the mountain, something from santero, and also [something] from the spirits of the mountain. That's where the Spanish and the African enter our Indian custom. But the cruza'o like me has the force of nature as base, which is the Indian element, and also powers from saints, because the saints came over here with the Christians and they adapted and we know how to invoke them like other spirits. It means that it is mixed, but it is how our way has continued to manifest itself—then the cruza'o has to handle it all.

The spirituality of the mountain, the respect and tenderness for the Earth, is more from nature, and our altar is natural; it's what comes to us from the mountain and it's one of the most Cuban things that have mixed.

My Indian prayer belongs to me. That's mine with certainty. And when I say mass, it's not a Catholic Mass, but a prayer we have. For me, a mass is praying, praying, praying; that's ceremony. And in it there's everything, and all power comes from it. That's difficult to explain, but the current we have comes to me from above, and in that I see, I find answers. When I see them, they can come all together or they can come separate, but spirits and santos, ceiba, *majá* or *guaorao,* Virgen del Cobre, and the Mother of Waters.

So it's the same if you do a mass or a *bembé* for a dead person, because people ask for those ceremonies.[68] From the bembé you take all the candles from the ceremony and you light nine candles—nine candles to make an altar and pray—and after that everyone gets up and lifts that altar toward glory and I have a spiritual chant that says, "Glorious, glorious glory / may they take you to glory / to the higher level, nicely kneeling on your knees / glorious, glorious mother, glorious, glorious glory / may they take you to glory, nicely bent at your knees." This is when you are going to raise a spirit in a mass, in a prayer.

From the Catholic we use the Lord's Prayer a lot, the Creed, the Hail Mary, and other prayers. But then we have the bembé, which as you know is African, and we look for the drum to play. We use the Indian drum. The bembé is like ceremony, almost a culture, and it's beautiful. I make a bembé every year on December 17. For a bembé I sacrifice a goat, as well as a pig, a hen, and a turkey. A bembé is one of the most beautiful celebrations done in the mountain, because it's about sweets, eating fruit, all kinds of fruit, candy, all kinds. That's bembé, which is for San Lázaro, for Elegguá, and for Changó.[69] And that's the African in our ways, and well, all of that is culture from the mountain, culture from the guajiro. It is inside what we have among Indios from La Caridad.

Really, that's what I carry as my religion. It's not evangelical or anything, but cruza'o, (blended) and that's our religion, which collaborates with the natural world of the mountain. Our grandparents

68 Bembé: sacred all-night ceremony dedicated to San Lázaro (del pueblo), celebrated the night of December 17.

69 Elegguá and Changó are Afrocuban deities. Elegguá sustains a privileged position as owner of roads and pathways, whch he can open or close. He signals the crossroads and in various ways is the owner of the future. Changó is an ancient warrior converted as a god, present in nearly all legend, and coupled in Cuba with the Catholic saint Santa Barbara.

recognized those powers well. Because they knew, they noticed. They'd use a coconut. They set up a güira, set up a coconut over there, leaving something for the powers of the Earth. Or they set up a little jar with sand for San Lázaro, to help us build. If not, one sets up a mug with a pot and a ribbon around it so people don't curse our plants with evil eye. Things are buried; the good is buried seeking harmony, peace, protection. That's what they used to do and that's how it's done. Those are customs from long ago, but people still hold that belief very much from the mountain. We have that belief and among us we see its result.

We also have our God of Clay, (*barro*) part of Mother Earth, because he is the one who sanctifies animals, sanctifies people, sanctifies fields. I do spiritual work on my field so that the plants grow. Look, right now I have something placed in a tomato field that represents me, and I say to that field with its soil, "Gather everything good and bring it to my meadow so the plants can grow." Anyway, that's a belief people had, but no one taught it to me; I went to it, and it is mine. We do spiritual work on our plants so that people who might want to come and steal it away can see that it says not to take anything here. I had a very big cucumber field, I harvested plenty cucumbers to eat, but the students' path to the school is right next to it. The children took my cucumbers. So I set up that bull's horn there, with my prayer, and from then on no one else took anything, because they respected it. They'd say, "Look, beware of what's there," and they'd go by and take nothing.

Nowadays there's less respect because today people are more civilized. They don't think as much about the things of the elders. But I do, I respect all that. Our way has its civilization as well, I believe.

People come looking for me

People come looking for me. Since I was young I've been healing my people. There are many things, many problems. Very good works have been done in the region, but there is still much to do. They come looking for me to do a mass, for my prayer, for my knowledge, for one thing or another.

Mine is not a Catholic Mass, a priest's mass, but rather our own ceremony, the ceremony I hold. My way is wanting to heal and to be granted the current (corrientazo), so the person is healed. What is mine is to be able to do it. That's why my children tell me I should keep a journal with the people who have found their healing here.

I healed a lady who was going to have surgery because of a fibroma she grew and she already had the doctor's appointment for the surgery. I worked on her and she had a discharge. It came out. I did the same with seven others; I prescribed purgatives and all seven didn't need surgery. Seven women didn't have to have surgery because I prescribed three purgatives from the baked güira juice to dissolve what they had inside. A cigar butt, *comején* [termite], and kitchen soot—that's a good purgative.[70]

Kitchen soot from where we cook with firewood and from the house made of guano creates a smoke sticking to the guano, which makes a sort of glue, and we cut three little pieces of that soot and mix it with the comején and the cigar butt. Comején is a little creature that grows in the wild. It has a large house. And it has medicine. I prescribed this purgative to the ladies with fibroma and they discharged so much blood. Anyone drinking that purgative has discharged everything they had inside. The women discharged the inflammation because they told me so later. They brought me the proof and said, "Look, this is all that I discharged." And I said, "Look, now go to the doctor." And they've gone to see the doctor who says, "And what did you do, what did you take? Because look, now you have nothing anymore."

This has happened to me. I don't exaggerate. They are cures that come to me. What I know about our plants, well, what our elders told us, and the dreams, the premonitions. I go to my spiritual center, to my altar. There I have the Virgen del Cobre, the moon, the Earth, and the sun. I also have Jesus Christ, San Lázaro, a Warrior, San Miguel. I have the water-spraying branch. I stand firmly and ask them for help, "Come on, let us heal." Then I tell San Rafael, the divine doctor, "Let us heal this woman." I ask him to take the pain away and call out like that. I see the results that can be obtained because many didn't have to have surgery and they love me.

During the ceremony the idea of what needs to be done comes to me, and I wonder: since I'm not a doctor, why does it enter my mind that this is good for that? And I say, it's nature, the spirits, including my own, who are indicating it to me, and the santos. The mind reacts.

---

70 Güira: *Crescencia cujete*. Comején: *Nasutitermes costalis*.

Look, I had a splitting headache and a high fever. That was on a Sunday and I felt the hand. I'm asleep with a headache and fever, a cold, and I feel a hand and a voice tells me, "Listen." That's what I think I heard. I thought it was my spouse I had lying next to me who said "listen," but I check and I see no one and she's asleep. And then I hear this: "Listen, you have the medicine to heal right here, a guarano leaf. Put three pieces into some tea." I concentrated in my bed and said, "Oh!" and I saw the guarano plant. Lying there I saw the guarano plant, where it is in the hills; I saw it right there before me and I quickly went there. "Where are you going?" my wife asked. I said, "Come on, put up a tin can or a pot for me." I went and picked a leaf of that guarano, *zap*, and I drank a glass. I drank it three times that day and the headache and the fever were gone.

That's why I say that tree has power or he who gave it to me is power also. Because it's the plant and the spirit who bring it to you. Some are Indian spirits from the mountain, some are santos, and others some are of the *abacuás* (Afrocuban religion). But mine is the way of the mountain and that's why I tell you I take such care of the beauty of it, because I've taken so much medicine. I've already taken so much medicine from trees and more trees and from the little plants the Earth offers that I have to love her, I have to like her—she is my mother and that's my everything. My children go and pick this powerful little plant and for me that plant is a doctor because of its healing power.

I had a niece who for many years had a heart murmur. Many doctors saw her. All said she was dying. She was hospitalized in Santiago, she was in Guantánamo, she was everywhere around here, and I said to my sister, "Look, give her *hinojo* [fennel] juice."[71] And she was sick, but she lived another ten years with what she drank, which cleansed her. At the end she caught one of those bad colds that brought her down. But, after having been gravely ill, she did enjoy life and lasted quite a long time with that medicine. She died, but everyone, we all have to die. She got well, she enjoyed life, and that's why I say that sometimes I lie down and think there are so many things in life, but nothing like the mountain.

I see the mountain above giving me so many beautiful things and I have so much love for her that, as I said yesterday, I love her

71 Jigüe, Lysiloma, Latisiliqua (L.) Benth

so much it's hard for me to scratch her, to open her up, but I have to do it to give us mutual strength, for us to live from her. It's like that, unfortunately, that a plant has to be cut down, because it might be in the way or we have to do something else. But it must hurt, because that plant is useful, that plant is a human, it's a being, that plant feels . . . they have so many powers and give so many fruits that animals as well as humans eat. My way is communication with plants, because those have a lot of power.

I take shelter in everything

In the spiritual world I take shelter in everything. I don't ponder whether I take from the Catholic, or from the African, because my foundation is the Earth. I dream about something and I do my ceremony. Not by someone telling me, but rather myself as I see fit with my own experience on my own account. And it might happen that I feel like doing a bembé, and I make a bembé for a saint. But, I have nothing to do with another santero coming and doing anything. I think to myself, "That cat, just what is it he can do for me?"

No, my way is nature, natural, and, well, I don't know if we'll put it down as my Indio way or as my own characteristic stemming from my own intelligence, or maybe a talent or my knowledge, or from what my people give to me; I don't know where that which comes to me originates from, because where does a dream come from? But it comes to me and then it comes from a tree, from a plant, or from the earth, or from other regions—for example, the chants and the santos carry and bring us much, as well as from the African tradition.

For example, Elegguá. What is Elegguá? It's an entity. And I know that and I respect it. And it comes to me. And that's certainly African. I have an Elegguá placed at the door to watch the house, but I do it with my system; I don't do it with someone else's system. I do it because Elegguá is here in Cuba. He looks for me and, if he offers himself, I respect him. My system is my own, but I feel respect and in the spiritual realm I take shelter in everything.

To all santeros who go to a bembé over by my altar, I say, "My way is different from yours. My way is natural, a thing of nature that comes to guide me." I tell them, "Nature is right there," and they sometimes say, "Come to my house." But I don't, no, no, no, I follow the natural way. I believe I have a friend in that tree; I have a friend and it's a friend of all these people of mine. I don't allow anyone to close my eyes and tell me, "Give me a little from here." No. That doesn't work for me. Neither do I charge anyone for preparing a bottle.

I don't charge for spiritual work, including the work it takes me to fix a bottle.[72] I don't charge anyone anything. I am humanitarian. And I shouldn't charge, because I'm humanitarian and if I charge you for this water, for saying a Lord's Prayer and a Creed, I'm taking your money, I'm taking your sweat, your children's food; no, no. Nature gave you that water. Man gave it to you, too, since he carries it over here, but it does originate from nature. How am I going to charge you for it? It cannot be because then I'm no longer humanitarian, I don't feel like a natural, I don't even feel like a man, an elder, or a woman, because we all feel the same. You get a headache and I know what it is like—I can feel it, too, when I'm in pain, although of course we all feel in different ways.

I'm not saying people don't help me. But without charging. One should never charge for spiritual work, because it's not yours. It's a gift you've been given. I work without charging, singing and working all night long, spending the night doing some work for a sick person. Working in two rounds, a choir here, another there, leading, "Sing this song, like this, like that." You do the work, but you don't charge. Of course people help us. Those of us who are in this trade, they help us.

Chiragua, Indian entity

I have a song. I don't know where it came to me. I have a song that goes, "The white scarf that came from the sky with Chiragua's crown." Chiragua is very much our own. Chiragua is very ancient. It's an Indian word. I saw Chiragua in a dream. And I heard his chant. I heard that saint in my dream singing about Chiragua's crown. And I know Chiragua is an Indigenous spirit, because I heard him singing in my dream.

My children ask me, "And where do so many songs emerge from?" They aren't mine. I just hear the ancestors singing them in my dreams, our grandparents' grandparents.

72 "Fix the bottle," or *hacer botella*, refers to the mix of plant medicine prescribed and made in a bottle for the afflicted person.

"With Chiragua's crown." He was an ancient cacique who descended from heaven with a white scarf and with Chiragua's crown. I did some work on a young woman from a rural place in the interior, who came with something that had her hospitalized and the doctors couldn't find anything. So they brought her to me. They were issues of pained spirit and I sang her that song. I sang her those chants and people asked, "And what of those old chants?"

To get into these and other songs, we're already in religion, we are in what is my religion. For example, "Lorisa, where do you come from?" is another kind of song. Anyway, there is so much I have in my mind that it escapes me. But [there is] "Lorisa, where do you come from? / from glory over there, from the rivers." So those songs mention the spirit's water, which is the Mother of Waters. "Lorisa from over there from the rivers. . ." and many songs are like that. I forget things now that you have asked me. And now those songs slip through my mind a bit, but oh well, they'll return.

The Enchantment is grand

I am asked about the enchantment. I would tell you that there are so many enchantments I wonder which one to start on.

I will start with the Earth, which is the most beautiful and grandest woman there is, most pretty and beautiful, where water runs through the canyons and you hear her rumbling, so attractive to a man, and you see that where the rivulets grow there runs a strand in the hair of the mother of the Earth.

Then the air that you breathe, which makes our life breathing, how constant that air is for us the human beings . . .

The greatness of the sun . . . what can I say. You either feel his greatness for yourself, or you are not capable of knowing it.

The moon is also a mother. She is grandmother. She is the water; she is the lantern.

The stars we call *lucero* [bright light]. When you see the night opened wide, you say, "Oh, what beauty," that moon of brightness, that sun that heats us. Look, if I just think about it, it is a poem . . .

It is all of that. That is the grandest tradition of life. From the making of the human being, the first human being had to count on all that naturalness. We leave but that stays, that stays here, and that's why we have to protect here, we have to protect all of that beauty that nature gave us so that we human beings can live through many long years . . .

When I travel down a path, from La Ranchería to Tames, and come upon the fields, I see the beautiful cane flowing in the wind, heated by the sun; I say to myself: who was it that gave us the Earth? Yes, the human may plant her, but it is the Earth that is maintaining us, the sugar as is born is really hers, that refreshment that we drink, and all the sweetness that we get belongs to the Earth and that is why we have to love her and more. Of all the beauty of life, the grandest is the Earth.

The Mother Earth

My grandmother would tell me, "We are the mothers of the children we birth, but the mother of everything is the Mother Earth, because without her no one can live." My grandfather would tell me, "Truly she is one we have to protect; she is the one that gives all the food that there is."

And look at all that is being done now on the Earth. There are so many—millions—of people who don't appreciate the richness that they have; they don't appreciate the Mother Earth. Because where does the richness come from? If they pulverize stone to find gold, to make spoons or pots and pans—that is all from the Earth. Copper, where does it come from? From the Earth they make the truck, they make the airplane, and now the people don't remember the Mother Earth, but that's where all steel comes from, from the Earth.

That tradition of the old people I carry in my heart. I tell my children, "I am going for a walk in the woods." The sons tell me, "Old man, don't go into the woods all alone." But I say, "I am going to go implore the Mother Earth."

One day I asked some artists, "Why don't we the artists all commit to the Mother Earth?" They were visiting La Ranchería. I told them, "Tell for me to all the Cuban artists that they each should plant on the Moher Earth, even one seed of whatever tree, of whatever fruit they like to eat, to fortify her, because without her, there is no singer, without her, there is not guitarist, nor a piano, because the piano is made by way of the earth, with metal and lead, whatever is needed, is from the earth. A guitar, too, is a wooden box with a neck. We cut it and get a guitar; it gives us a *tres*, (Cuban three-string guitar), but we have to thank our mother for this, thank the Mother Earth. And to anyone walking today I make this call, that he should implore her every day, and to give her thanks. Because the Earth hears, she does.

I say there are three things equal to all nature. The Earth, the Moon, and the Woman—these are three things. We are not to compare the power of the Earth or the moon to the power of a woman, because those beings are very powerful. But without woman, there is no world. I see a woman carrying a child in her womb and I tell a woman, "You are the same as the Earth, because Earth gives us our food—everything, everything, grains, fruits, tubers, animals—the Earth gives all sustenance to animals, to the human being, to the chicken, the small birds, the big birds, because the Earth gives fruits for a bird to eat up there, and to all critters of the Earth, to everything, everything. But for the Earth, they would not live. We have to think about that.

The new generation has to ask, "Why are we here?" For the Earth, through my mother I eat ñame malanga, bean, rice sweets. Well, who gives us the sweets? The Earth. The honeybee manufactures it in the field, and in the sugarmills they produce it with steel, but if you don't plant the cane, there is no sugar; if there is no earth, there is no sugar, and that land came so that the world will be complete, and so that we protect her, so that we adore her, so that we plant on her. Here we are standing on a floor, but we see that plant full of flowers and all the plants that grow from a trunk that has roots in the earth, which is introduced into the tree. A chicken, a pig, a horse, an oxen—what are they standing upon? On the earth, eating and rolling in it, on the earth. The earth, soil, is clean. You see many that say, no, because I might get muddied. I say, well, get muddied, because the Mother Earth itself tells you to get muddy. Water comes and refreshes the earth, and I stand over there, in my place, and I say, boys, that grass that you see so green, that is fed by the Earth.

An ignorant man passed by my house, a chief of something, and he said to me thus: "Ignoramus, why do you have such muddy, grassy patios?"

"You are the ignorant one," I tell him. "And what would be the image of the Earth? It is precisely that little grass that sustains her; that is her image." I walk over that little grass of the Earth and I feel her in my legs, a softness. But if I walk on a highway, or in the town, where all is concrete and concrete, my legs tire on me. I sit on a stone, in a big garden or in the road to look over the grasses as it plays with the Earth and I tell her, "If but for you, how would I live?"

There are people who stop and ask, Panchito, you think such and such crop will grow here?

I say, You have to test. Test with one small bean. If it sprouts well and grows to good fruition, then that land is good for that. There are various colors of earth, because earth is the same as people, who also have four colors and also the birds, the animals. The earth is the same: white, yellow, red, black. Eh? And each has its substance.

All my life, I rolled it with the earth. My luck is in the land, because without her, we are nobody.

Look, the thing is that the Earth responds to me. In the cábula, she responds. Ah, when I am going to plant, I take notice of the Earth, and in nature, in the land, the water. There I say, Mother of Waters, I am going to plant. There is a need for rain. The boy was to plant *boniato* [yam], and he was preparing the soil and I was cutting vines when the sky suddenly clouded. "Look at this cábula!" Water fell that day and two days later we were planting boniato. I would ask my grandfather, "Abuelo, and just what is cábula?" He says, "Boy, cábula is a notice from the Earth."

The four directions

I also work with the four directions. When I see a patient who is very ill, I take a machete and stand up and draw the four directions, and I go and say, "Look, bury this victorious sword in a corner and ask the four directions to take power over that victorious sword in order to cure this sick person." Then I draw a cross of water and cleave the machete, and I ask the people to go there at twelve at night, and those who do I instruct on how to take care of it, how to take away the illness. That is a system that I follow.

The santo or the god or deity always has a cross. And this I dedicate to the Cross of Jesus Christ. That's also called a Cross of Water. Well, what I draw around that cross is a circle and a cross of ashes, with two candles forming a cross. And I place the person I'm going to cleanse in the center and cleanse her; the ashes I throw into the alcohol start burning and I stir it so she receives some heat. I say "some heat" because here we all are children of the heat, of the cold, of the medium heat—all are motivated by it—and what I mean is that human beings feel all there is because we benefit from everything. Because heat is beneficial, and the change. You sweat through the pores and when you are very hot; the cold comes and refreshes you. Well, that's healthy for man or human being or anyone.

The ashes used are not just any ashes, but the ones from the stove in your home, precisely the ashes coming from those sticks you pile up in the burner to cook with. Those are the ashes you take for this act. I work on you, along with my assistant. We work on you in this act, on the floor and with those ashes. In that circle of protection, when I cast a cross over by that door, I throw a drop here by the other door. I make three crosses for you and I light them up and have you cross over there by that cross of ashes.

And if we have to use a white hen's egg, we use it. For that you take an egg and dedicate it to Santa Clara the powerful, and looking toward the sun you hold up the egg and say, "Santa Clara, this egg is to cure Francisco Ramírez." That's how it's offered. Then I have you put that egg into a bucket of water and place it in the sun, and by noon I tell you to bathe with that water. The egg you take without breaking as if it were soap and pass it over here and there, and then you place that egg in a little corner until the next day. Be it four baths, be it five, that egg has to be there and within five days of having the egg you do the cleansing bath, and after that you go and take that egg now filled with all your wrongs and break it at a three-way or four-way crossroad so the bad influence, you see, so the bad influence wears out in the world. Those are the four points. The four directions of the world grab the force or release it. It belongs there.

The four directions are also seen in the raising of a bohío. When making a bohío you watch its four directions to build it. They used to sprinkle water as if to baptize the home before moving in. An old man used to say, "Listen girl, sprinkle water on the house before moving in. May no one be able to get into that pagan home." They called it "pagan" because it still belonged to the forest. And if anyone doubted it, the old man said, "No way, we need to find a preacher to pray to the four corners."

Notice that they'd take a little rompezaragüey or basil branch and a glass of water, and they'd go to every pillar and pray. They'd go to the four points of the four corners and they sprinkled the door and thus the house was sanctified, baptized. That's done to the east, to the west, to the north, and to the south.

Nowadays it's rarely done, but it was back then. Those methods were followed back then and they were experiences one had. And it's something that ought to be done. I do it; I place my four little points there. If I build a house, for example, I say I'm going to baptize the house before moving in and *bam, bam, bam* I cast my cross of directions and I know how to pray.

Also sometimes if a youngster gets sick they say, "Sanctify him, because he has the evil eye." And the four directions are used for this. In a hen coop [it is] the same. And when we're going to take a bird or an animal out of the patio, out of the batey, if it's a hen I pull three feathers and bury them in the soil. Also if it's a goat, I pull a little hair; if it's a pig, I pull some hair from its back and I plant it, so that it'll never run out. One leaves but within a short time the Earth sends another. And that's a very old custom of ours. I follow all those customs—ancient customs and all from my grandparents. They are customs that are no longer followed with precision because our young have been around a lot, because young people have studied a lot, but not I. It doesn't leave me and I continue to follow it. I tell them, "Butcher the hen and bury the feathers, so you never run out. And pray to the four directions, the cross that belonged to us."

Towards the four directions—planting

Myself, when I am preparing to travel, I take stock of my course. The day of my departure to arrive here, I went to the four directions. I said, the sea is this way, the sun emerges here and posits there, the moon comes out here and lays down there. I look at them, I tell them, come with me; come, let's fly. I am like that; I always make my prayers.

The four cardinal points and the seven natural potencies. These are the earth, the sun, the moon, stars and all brightness, water, sky. Water is [also] the cloud, in the seven natural potencies that we all share, everyone. We need to work with them because they give us and can be approached. Because I take a little water and I give it to this little plant, and I will watch it grow. If no water falls, there is no food for the animals, nor for *cristianos* [Christians], for nothing. Water is a natural potency.

The sea. How many fish are in the ocean?

What the earth has given us, we have not known how to make good use of. I tell the chiefs and the not-chiefs, and to those who turn up the earth, who burn or even when they plow her. I tell them, "Listen, what nature gives us must be respected."

The Earth gives us guava, *anon* [custard apple], everything, everything. Guava, *aguacate* [avocado], *mamey* , [mammee (sapote)] *ciruela* [plum]—how rich are those fruits.

I propose that, if we could crop it all—for example, through all the provinces, not only Camagüey or Guantánamo, but all, in all the highways—to plant prune trees, from Guantánamo to Pinar de Rio, to have all of the highways planted in guava, in mango, anon, so that all travelers can say, oh, look at all the fruit, let's eat some.

Not like now. They burn all of it, so that they don't have to care for it.

I propose how rich it would be to do so all the way to Havana, or to Pinar de Rio, a *cordillera* [mountain range] of fruit trees from one end to the other, not harmful to the actual highways. That would be very beautiful, would be most wonderful, to travel looking at this. In our region, the people could travel the roads and paths and see fruit in abundance.

The Fruit drops and dries? New seed spreads; new seed is born.

As I mentioned, if all of our artists united and each planted a squash, each planted some beans, Cuba would be the richest country. Our old people were that way, singing as they planted.

I have various trees that I approach. I go and embrace him and say, give me your bark; give me your leaf to drink. The *guama* [ice-cream bean] tree, is an example, which grows not far over there. People come to ask me, Panchito, what's good for the kidneys? I say, dig around the root of that guama tree, take a piece of his root, drink from it, plant some more, drink from it. It [the tea] makes you urinate in quantity, and it doesn't harm you. Look, in our earlier days, we were surrounded by natural resources. Let's take the advantage and teach our new generation, because you will leave and I will leave, but they remain.

And that new generation, your son, my son, will know that that tree is good medicine. In that way, we help science develop.

Family spirits

I hold mass for spirits like my grandfather's. Not long ago I held a family mass for my mother, for my aunts and uncles, for my grandparents. I light nine candles and have all the people praying—those who are family—and I'm there with them.

In my own tradition you don't offer food to spirits; to saints, yes, and to trees, to plants, to the mountain yes, but not to family spirits because they are light. Since they are light we give them light, candle, fire.

But there is a family system many people follow, where they go and leave a little plate of food for their dead at the cemetery. I respect them, but I've never done it. For a saint or main spirit, yes, I've given food and butchered goats and pigs and I've taken the blood and offered it on a little plate to the saints and also to Babalú and to Changó, to Eleggua, to Oggún. The important thing is to show respect and that's my system, where you respect every spirit in existence on our Earth without any shame. Always respecting the natural spirits, the trees, the grove, the waters, the earth itself. So if I'm going to butcher a rooster, for example, I also put the rooster's blood under a tree, a guásima, an orange tree. It all lives in its own way and can help us.

It's the same with the grandparents, with all of them on the other side. We always have to remember them, thank them, talk with them so they feel good. And it's true that they can help us, too.

The ceiba, sacred tree

A short while ago they were talking to me about the ceiba, since I do work by its trunk.[73] I ask for its help because it's a huge tree, a very powerful tree. We ask it for help, it being so big, and offer a candle to it. It's an act of devotion we do. Close to where I live there are no ceibas, but there are very far away, and I've gone there and said, "Look, there's a person in danger, exposed to danger." I ask it for life. I go to a small and chubby little ceiba and that's the one I ask. The ceiba fills

73 Ceiba: *Ceiba Pendantra* (L.) *Gaertn*. In Cuba, the ceiba is considered "tree of peace" or "tree of life."

itself with sacred wind. It knows how to bless and has the power of lightning, the strength of lightning and thunder.

And be careful with that tree; for sure, you mustn't cut down a ceiba. I for one don't do it and if it belongs to the field, I leave it there. If I'm going to cultivate a field and there is a ceiba, I don't cut it, I leave it be, just like a sacred palm tree. Everything the ceiba has undertaken, as I say, is sacred, and we perform works by the ceiba. It's sacred, a very sacred tree. It's because the ceiba also has its mystery. It is strong for cleansing. When someone is cleansed under it, you ask that ceiba—since it's luxuriously leafy, since it's one of the largest and most mysterious trees in existence—you ask it for fortitude. Those strong trees have their charm. I walk by their foot and I ask them for faith, for faith to grant me resistance in my life, because I'm almost like them. Yes, I walk, I talk, I can throw. They don't throw and they don't talk with words, but they do through their touch. Yes, they are standing there by themselves, but they breathe and that's why I was telling you there is life, there is consciousness, and there is strength.

There are so many beautiful things in the wild that one can never forget the mountain. I'm one of those I'm telling you about, who never forget the mountain. Never in my life will I forget—only once I disappear. And that's how it is and I tell my children to stay here, to not live in town. And my children do say they follow my system in the mountain, under a tree with all the necessary breeze.

Because you see, in the city you have to have man's resources. If you don't have them, you burn up. Even an electric fan, important in the city, is man's effort, made by man. Now, the air here is nature made. Trees. That's the one I receive; it's the one that refreshes me. And, also, it is sacred.

Lying down with what the mountain tells you, the little sounds of the mountain give me life, fortitude. The crystalline water that surfaces from under the earth blesses me. Sometimes I walk through the hills and I say, "What a beautiful tree! What might be its strength?" I lean against it and it's like sitting in an armchair. I lean against a ceiba and other trees to make me feel strong like them and I leave satisfied.

My grandparents used to say the ceiba was sacred, that you couldn't cut it down, first because the *majá* [Cuban boa] makes its home in its nooks and crannies. The majá searches for the ceiba to save itself; the ceiba is protection, is its home, my grandfather said, and you always watch over the majá. And [second] because the ceiba's trunk, as big as it is, even though it's a tree with thorns, has bark that is at the same time very smooth.

The majá has its significance, too. The majá is a creature, we sometimes say a very silly creature, because if you don't bump into it, it isn't aggressive, but if you do, it is and it should be left alone. The majá is useful. Majá lard is useful and many take it for nourishment. The lard is good for many things; for example, for any chest tightness you might have, you use majá lard. Another benefit of the majá is that it eats mice, and that's very useful. Many among our people who see a majá in a home won't kill it for the mystery it has of keeping away mice, keeping away anything. Now, when it's going to eat a chicken, then you certainly have to scare it off or kill it. I respect the majá; I don't kill it.

The guásima

The guásima is a benefit for humanity.[74] It's useful against human and animal parasites, [it's useful] for animal fodder, the seeds; the ram, the cow and the horse eat and make good use of the leaves; the wood is good for yokes, for boards, for musical instruments, for baseball bats. Imagine how important it is that in medicine the guásima is the best for parasites. And when you are so constipated that you can't defecate, take some guásima bark—when it's slimy it's very good. The guásima is also used in spiritual work. Not long ago I recommended some for a person who came to my house. I said, "Take an egg and bathe with it and break it in the middle, in the center of a guásima's forked branch." And she did it and said, "Well, you know this remedy was magnificent for me." And that's the power of the guásima. It's the power that tree gave me and her. As I'm telling you, I have faith; I have faith in it and that's why it responds to me. For me, the guásima has power, and it has fortitude because it's very medicinal. It hasn't been discovered yet, not even by science experts, what a tree like the guásima truly means for human beings. You take a bit of guásima and you take a bit of soil and you mix it and with that you make glue. The guásima is eaten by cattle. Its bark is useful, its wood is useful, all is useful, everything of the guásima is useful. Imagine that in the bateys in the mountain there

74 Guásima: *Guazuma tomentosa.*

always are guásimas, and the mountain creatures come to them, be it jutía, birds, and *chipojos* [anoles] and all that, to enjoy themselves.

The song of the caguairán

There is a tree called the *caguairán*.[75] Its wood is hard, good for houses, but it has a liquid inside that we call wine and we used to drink it. You'd cut a very large caguairán and when you reached its heart, the center of the tree, it released some water, which we tree cutters collected. When drinking it, it made us drunk. It's like drinking wine. That tree hardly exists anymore. It exists in very few places. It's an extremely strong wood. It's good to make pylons. The bark is good to drink as a tea, the leaves are prescribed for baths, and the bark and wood are put in ordinary water and it turns a reddish color. And it's good for cleansing blood. That tree was abundant here at La Ranchería, but in the past there were lots of fires and it was very much destroyed. But those are timbers we have in our forest and today they are being rescued. There are places where they are being rescued because that tree was almost lost to the mountains.

So I have a chant for that tree. It's one of those chants that surges out of me, like a sung saint that comes to me. I heard that saint's chant in a dream and it told me it adheres itself to that tree. That's what happens: things interconnect in nature and sometimes they reveal themselves. As I was saying, my way is a matter of dreaming. If I dream it I take it for real. So, no one sings that chant except for me and it goes: "Oh, what a strong tree, San Miguel / the strongest tree caguairán / Oh, what a strong tree, San Miguel / oh, what a strong tree, San Miguel / oh, strongest tree caguairán / oh, may you do good, San Miguel / what a strong tree San Miguel, oh, what a strong tree, San Miguel / oh, what a strong tree San Miguel, oh, what a strong tree caguairán / oh, what a strong tree San Miguel."

I combined that in works I was doing with many sick people and I'd take and prescribe remedies from that tree, because it's a strong tree—not easy for the ax and the machete to cut. Everything made of iron is dented by that tree and it also gives us that delicious little water. So through that dream I noticed that this is the best tree to do work with.

---

75 Caguairán: *Guibourtia hymenaeifolia.* It is a tall tree with very dense wood.

I ask from the very tree to get its strength, since it's a tree that blunts everything that hits it and knows how to defend itself. And what I do, it just comes to me, is ask for the strength of that tree, ask in the name of San Miguel, a saint who has tremendous fortitude as well, and who has aligned himself to the natural force of the caguairán. [With] that saint, when I'm working like that and singing, I feel despair and hot inside. Right now while singing, just from showing it to you, I am hot, a heat that rises, rises all the way up and then drops. So I have faith in the tree and in the saint and I use them together, because I feel it, I feel that heat, I feel the body ignited, I sweat. See, I'm sweating when I sing that song to the being. That's how I find my way around within my medicine—with dreams and with my body, with my body.

The jagüey and the branches of three powers

For making my prayers I also use the jagüey.[76] I now have a very big jagüey tree. I go there and sanctify people and they bathe underneath the jagüey. And I do that with branches of three powers, of *providencia* [Providence], of siguaraya [Trichilia havanensis] of rompezaragüey. They take a bath of all that under the tree to ask from it. I have that tree dedicated to the Seven Powers of the Road, so they take over that tree, so they take over the illness that's with a sick person, to take it away. That tree has the power to take away illness. I ask the Saint of the Mountain, the Blessed Saints. I take the same three powers. From plants we have at home I take the branches that are medicinal. There are the saints, the spirits of the mountain, and the house saints who are the blessed ones. The spirits from the mountain are the most natural ones. They also call them pagan saints because they are in the wild. In those of the mountain resides the connection to our Indian way, in that belief. The trees have spirit and they recognize you if you greet them with your prayer, respect, and sincerity. It's in those trees where the energies lie, because from the trees comes a lot of medicinal energy; their power comes from their natural spirit. It's a fact.

76 Jagüey, *Ficus trigonata* (L.).

The majá of the jagüey

In a place called Ceiba, something very sacred happened to me, one of those gifts that come from nature when you're sincere with her. There was a little spot that I discovered some ten years ago where a jagüey tree stands, and when it's noon with the sun shining and there are no clouds, water falls from it.

That was a great mystery for me, that jagüey. You'd walk underneath it and you'd get wet, but try to collect that water and you couldn't. I went to collect the water that came down on a plate at noon and not a single drop fell on the plate. I set up a little tripod of sticks, I placed a plate on it, I set up a stick to tie the vessel to a pot to bring that little water to the laboratory, and I couldn't get a single spoonful of water. The mystery was that I'd go there when I was very hot, so the water would fall on me because it's fresh water and I'd come out with wet hair and feeling very blessed.

I had a *madrina* (godmother), who has died, but she'd tell me, "Godson, you're bold. How do you get under that weeping tree, only God knows what's there. You get under that tree?" I'd answer, "No, godmother, I'm natural just like the soil, like this tree. Water runs from me and I'm the same and it won't hurt me because I love it."

"That's your thing," she'd say. "I've been living here for many years," she said, "and here everyone walked under that tree and no one got wet, and when you arrived here it gets everyone passing by wet." She said, "That's yours, Panchito, your spirits that come here to enjoy and get you wet, bathe you."

That one is a mysterious tree for me. To refresh me it gave me water, but it didn't let me take it away. There were two young men and a veterinary technician who went to see it. The veterinary technician wanted to know what was happening, if a natural phenomenon was causing this. He climbed up to collect leaves off the tree, and inside a tree fork of the jagüey, very high up, a great majá appeared, and the man jumped off the tree as fast as he could.

The young men made the majá flee by throwing rocks. But after that majá left, the jagüey never rained again. I have walked by and the same young men tell me, "The majá left and it doesn't rain anymore." I say to them, "It left because you were violent to it."

"*Madrina*," I'd say to her, "But how is it I can have that power?" And she said, "No, son, understand, it's not you, but what you have, what you have makes all of that happen. Don't you pray, don't you ask the nature of the world, the Lord and Christ?"

"I do, for health of the whole world, for life for the whole world."

"Well, the tree has what's natural, just like a god, and it granted you that water."

And imagine, the majá left and the water left. That's why for me the majá has a spiritual sense. It has the power of a mediator, my father used to say. The majá and lightning, that snake has its current; it has its current.

A man here in the mountain used the majá as a sort of backup for his farm. He was an Englishman whom I knew and he always had a big majá on the farm and the majá lived in a ceiba, also big. That great majá almost whistled like a rooster. Because big majás whistle, you know; they almost whistle like roosters. But it so happened that one day the Englishman's majá went downstream and left. They say it got offended, crowed three times toward the Englishman, took off down the ravine, and disappeared. Shortly after, the farm went bankrupt and the man died from a shot. What happened? Why did he die? Because all the power the farm had had left.

I don't kill a majá, I feel spiritual respect for it. There are those who kill it; I don't. There are also those who breed them. I take care of them in the mountain, but I don't breed them. Some do. Our grandparents left it alone, because if you don't bother it, it won't hurt you. On the contrary, it can benefit you since it eats rats that come up to your house to eat little chicks. They can also cause harm if they eat the chicken, or the hen, kill the lamb, or the pig, because they are wild animals. When the majá is very hungry and doesn't find food in the forest, it'll eat the animals in the bohío. Anyhow, when it begins doing harm, you remove it.

Cedar and cleansing smudge

I talked to you about the caguairán. There are many medicinal trees, many trees for the benefit of man and humanity at large. The jigüe tree[77] for wood, and the cedar[78] for medicine, and all their barks are useful. But cedar is the best for smudging.

All those trees have their mystery, because I tell my children, "Trees don't talk, but they feel when you cut them because they breathe and search for water with their little roots. And they know how to search. With their little roots and their canopy, that's it, the tree searches and searches for water. And when there is lots of forest it rains more."

That's why wood and forests in Cuba are not allowed to be burned. So it may always rain, and if we don't obey, if we kill all those groves, what are we doing? Because for me, they are living beings, exactly like us. And they have their talent; they have their duty. If they must be used, good, you ask for permission. You think and you ask permission—from the state and spiritual permission.

Cuba is already mountain everywhere, and it's becoming another new Cuba as we all take care of not setting fires, everyone taking care of their wood, planting. The state here has planted millions and millions of all kinds of trees and we also plant other important trees.

The guava (*guayaba*)

The guava is medicine.[79] It's good for diarrhea. The guava is strong for animals as well as for us cristianos.[80] It's a very tasty fruit. Just yesterday we ate guava and how tasty it was! Children go to the guava trees often, and at my place there are many, many trees, very many. We leave guavas for the spirits of the mountain; it's good to leave guavas; we offer them here at ceremonial celebrations. It used to be a wild fruit, of those that the mountain gives, and sometimes I take a guava and tell the youngsters in the field, "Don't cut that guava tree, because it's forbidden to cut guava, because it's a resource for humanity."

You are hungry and you go to a nicely full guava tree, and you've eaten. That's what that plant does for us. That's why we have to care for it because that tree gives humans strength and it is medicinal. The skin is medicinal; everything is useful.

---

77 Jigüe: *Lysiloma latisiliquum* (L.) Benth.; wild tamarind.

78 Cedro: *Cedrela odorata* (L.). Cedar is considered a medicinal sacrament in many Indigenous traditions, including Lakota, Haudenosaunee, Anishinaabe, and Maya, among others. Its smoke is used to cleanse sacred instruments.

79 Guava: *Psidium guajava* (L.).

80 As in medieval Spanish, and regional diction, Panchito interchanges "cristiano" with "human being."

It's not a big tree, but a tree that gives fruit. It gives us preserves, compote, juice; that's important and here we take care of it. We take care of it on every road you can see. Around here there is guava everywhere and no one gropes it. We have to take care of it, because it's food. A hungry wanderer takes three guavas and that's it. Eating guava, I'm on my way and back again. There are sour ones and some are sweet, and the sour ones are the ones for stomach disorders.

The sabina, the caoba, and other trees

Trees are our friends and neighbors. The *sabina* tree is a bit soft, but of very precious wood; however it also has almost disappeared because of fires. It's a tree that smells delicious; it has a very aromatic odor. You use a sliver to put in clothes and it gives a good smell. It's good for rheumatism.

There is the *caoba* [mahogany tree]. I have good caoba at my place, a caoba this fat, so I'm waiting a few more years, I've told the Forest Enterprise, to make myself a set of caoba furniture, which are very beautiful. Caoba is one of the most precious woods. I take care so those caobas standing over there won't burn. I have a few caobas on the farm and I care for them like children. They are trees that ought to be preciously taken care of.

Caoba bark is good for drinking. [If] I have a flu, I take some bark and boil it and I drink the tea. It soothes the cough. There is the *yagruma*,[81] I believe *yagruma* is an Indian word. Yagruma is a spongy wood, which means it's very soft, but we haven't given it enough attention yet, because yagruma has a lot to offer. Look, the bark is good for drinking it. It gives you appetite. You drink a little yagruma tea. You boil the bark in a bit of ordinary water and it turns rose-colored. Anyway, that bark makes you sort of hungry, so it sustains you, it gives you blood. Importantly, the leaves are eaten by cattle. I say we haven't given it enough attention because we still don't know well what nature is, how to better use nature and, what it gives us. Look, if we planted enough yagruma, when there is drought we could, I think, feed cattle with yagruma, because cattle eat it really well. And you take the leaves to use the little leaf shoot for a cold, for fever, for almost anything you

81 Yagruma: *Cecropia peltata* (L.).

might be feeling, for rheumatism; they're good remedies we know from the ancestors.

The only tree I know out of so many trees that is not useful and is toxic is the *yaba*.[82] And imagine that although it's malignant, it is useful for railroad ties and for house posts because it has a hard core, but it is toxic. That one poisons, because we've even done tests with ants. *Yaba* is an Indian word of ours, too. Yaba is quite toxic. No bugs perforate it nor peck it; birds hardly ever rest on those trees. Even the yaba smoke is harmful; it blinds you. It does have usefulness for houses, for railroad ties, and it has strength for carriages, for bolts, for many things, but that's the one we don't take for medicine, not for animals nor for anything else.

There is the *guao* [maiden plum] tree[83] which is toxic, but is also useful for medicine. You take its milk when you get *ojo de pescado*,[84] as they say when you get a little protrusion like that. You take and you spread the guao milk, but very carefully, because it's toxic. You have to test it on yourself first before applying it, because it causes swelling in many people. It doesn't work for me, for example. Guao I only use when very dry, for claves.[85] And it sounds very nice, *ta, taca, ta*, the guao. But it's toxic; it's dangerous. Even by walking under a guao tree you can develop a rash. You can become appallingly swollen, like burned.

There is another tree they call *juan pérez*. That one is definitely bad. It makes you look like a monster. It's a softish wood, almost useless. It's only good to cause harm. It's harmful because not even cattle eat it. Lots of trees exist in the world over there in the mountains.

The *ramón* [brushwood] tree has two types. [There is] one ramón that's used for making houses, setting up fences or railings, and the leaf is good for cows, for goats, for sheep, for pigs, for oxen—well, for all kinds of animals. Then there is another kind of ramón that has a broad leaf they call "going to live" and cattle can eat a lot of that one. Goats and horses eat it less, but when food is scarce you cut some going-to-live and sheep and cows eat it with pleasure. I find that important because you cut a bundle of branches from the going-to-live and put it out for a cow and that cow can sustain.

---

82 Yaba: *Andira inermis* (Sw.)..
83 Guao: *Comocladia dentata Jacq.*
84 "Fish eye," or a wart that protrudes on the skin.
85 Claves are high-pitched musical sticks that provide the key for a song.

The *acaná*[86] is also a tree that gets big and no longer grows in these mountains. There are no more. However, [from] that one you take the bark and it's good for stomach ulcers. It releases delicious, somewhat stained water, and it goes into your stomach and heals ulcers.

There's the oak for your teeth, good for a toothache. And also the *sabicú*[87] rinse is good for a bad tooth. And notice these are Indian words and those are timbers that when we see them I say, "Don't cut it down. Careful with that tree, since that one is used as mouth rinse for teeth."

The *maboa*—that's an Indian word—knocks teeth out as well. Maboa is a tree that releases a lot of stained milk, so if a drop falls on you it sticks on. Maboa is used for teeth, to pull teeth. You take a mouthful of that water and the bad tooth falls out. The mountain has so many things.

The *guamá*[88] is a real tree that's very common in La Ranchería. The people from Santiago are crazy about me taking a guamá tree to plant it in El Caney in the Atheneum over in Santiago where there are many Indios, too, as a memory to the Cacique Guamá. There are lots of those trees in La Ranchería. They are good for spiritual cleansing baths. The guamá root is used for kidneys. And that wood is felled during the waning moon because otherwise it gets infested with bugs. They come and perforate it if cut during any other moon. So it has to be felled during the waning moon. That tree produces a lot of water during the new moon.[89]

There is the *yamagua,*[90] which is very much indicated against hemorrhages. You cut yourself and you prepare a yamagua paste or make some tea or put the yamagua leaf on it and the blood is cut off. [If you have a] hemorrhage, you drink some tea. It's the best. That tree grows big like an oak. *Yamagua* is our word, and the Indian word as well. "Yama" is invoked; it's used in religion, too. It's a short form of *yamagua* because that prayer is to call the things of yamagua.

---

86 Acaná: *Manilkara albescens* (Griseb).
87 Sabicú: *Lysiloma sabicu* (L.) Benth.
88 Guamá: *Lonchocarpus domingensis* (Pers).
89 The cutting of trees and many instructions for planting the Taíno tuber crops in relation to the phases of the moon has aboriginal origins in Cuba and the Antilles.
90 Yamagua: *Guarea Trichilioides* (L.).

The cupey tree is San Aparicio

I have faith in another tree and take people to stand under it and have them light a candle underneath him, and that is the *cupey* [autograph tree].[91] That one is another sacred tree for me. The cupey, we were told, is a tree that communicates. Imagine that they write on its leaves. Even our Martí took the cupey leaves to write when there was no paper, and they wrote letters with those leaves.

That tree is not medicinal, but it has its mystery and I always use it for dedication in my things of nature, of religion. I cross it and I dedicate it to San Aparicio. I light a candle for a cupey root, for San Aparicio, in order to make all evil disappear and for the tree to cover us, because it grows very big, very abundant. It's very strong, that tree, and very valuable. They are trees one has faith in, as if they were one's god. I go to it and I like being with the tree. I enjoy myself under those trees. Their shade is spiritual. Sometimes I lie down when I go to one. I sit on its roots, I lie down a bit to rest, always at its base, and as I say I have faith in it.

That tree I hold as San Aparicio, because that's how I saw him and I feel he works, so that evil disappears and the good may come. And I burn my little candle for the cupey and ask it for fortitude since it's a very large tree, because I say, if you want to get shade under a tree, look for one that's leafy, big. Seeking that spiritual shade, I don't go and stand under a tree that doesn't give me any. I like that big, lush tree to give me fortitude and that's the faith I have in that kind of natural tree grove. I sit under the tree and it, collaborating with the air, moves its leaves to bring me good tidings of what nature is. Later, in my dream, I see once again that entire big grove, moving this way and that; the wind hits it and it moves. In my dreams I have flown and I have climbed on top of that grove, and I enjoy myself up there and learn things.

In my dreams I always go to where that tree is. I don't know what it might be. Perhaps it's something we feel only as Indios, or if this understanding spreads around our country, but as we lie down to sleep sometimes our spirit gets underway and goes toward those groves, looking for that tree spirit that has approached us.

---

91 Cupey: *Clusia Rosea* Jacq. Rebel Cubans would write (scratch) messages to each other on leaves of the Cupey.

That more caoba grow

Look how things are done here in La Ranchería. If I want to cut down a caoba to make some furniture, before I do it I first look up and I know it saddens me, but I have to cut it because I need it.[92] So, first I plead that on this earth I'm stepping upon, more caoba may come, that the little trees that are growing up become big, that they may grow, that for the caoba I'm going to cut hopefully another ten will rise just like that. It's the same with a palm tree. It weighs on me to cut it, but it was used to build the walls of houses, and since houses were bohíos that tree was felled a lot. I already said, "I won't cut any of the palm trees." Nowadays there are people who come here to the forest to cut trees. They go into the mountain not knowing and fell the tree, ignoring if it is needed, if that tree has to be there for something. They just go into the forests like that—but not I. I say to my people, "No, no, you can't go there. That needs to be respected. I myself, who belong here, don't go in there to cut down just like that not knowing, and now you are going to go in." To a few I've said, "Not over there, no—get out of my forest. Look, all of that forest is mine, I take care of it, do not cut trees there." And it's because they don't know. They need sticks for a fence and they cut down beautiful wood not used for that. I say, "You need a fence, good, ask me for it." I know where to go and cut the needed timber and not just chop it down like that—it's disrespectful.

We must care for the trees

For me, most important is that we have to take care of groves; otherwise it would be a desert if everything was cut down, if we don't take care of the grove, and of the little birds as well. All the animals natural to the mountain also must be protected. That's important, because they're also help for us humans, as they are part of us. They always told us that. Little animals, like plants as well as air and water, the earth itself, all of that are we. Since forever the earth are we, we turn into earth and from there we return. It protects us; it gives us life; it's like a relative; it's my family. When I talk about Indio, that's Indio.

About three days ago I met with five boys over at my place. They come to talk to me because I'm the one in charge of keeping watch for

92 Caoba: *Swietenia mahagoni;* West Indian mahogany.

the C.D.R.[93] The teacher brings them to me and he explains, "Look, Panchito, the boys here—there was a hummingbird nest and they broke its little eggs." [94] And I said, "And you didn't give them a couple of butt smacks." I had to let them know that those little birds need to be taken care of because that little bird sings beautifully and there's no nicer gift for humans than the singing of a little bird in the wild. And it's a being like us, like all animals and also the little plants. "Look," I said, "neither the *cartacuba* [Cuban tody] nor the *torolico* , who carry the national emblem, can be killed. And you are breaking the little eggs of the cartacubita." It carries the emblem of the Cuban flag, the torolico, the *mayito* [blackbird], and the cartacuba—those birds are national birds. The torolico has a cross on its tail and I don't allow that bird to be killed. The torolico grows a cross on its tail and it's the cross of the four directions and also of Jesus; that's the cross they put on Jesus and they put on the Taínos, and it's the same cross the Cubans are carrying upon them, too.[95]

Bee honey is the flowers' cream

Another thing we shouldn't forget is that the old Indios ate and we eat a lot of bee honey. That's extremely important. The numbers of beehives in Cuba are also greatly increasing. The Indio of the past, my grandparents said, lasted so long because he often ate bee honey. They sought beauty, what presented itself nicely, what offers itself with its own identity. Well, it's obvious. It's nature who is a great lady, a scientist. Listen, because we already knew that. We knew it and we know. Earth Mother, nature, is intelligent. And honey for us Indios of the mountains is a chemistry the bees extract, which is the cream of flowers, and I say, I repeat it to my people, "Look, eat honey, because it comes from the flowers of trees." So in my opinion that offers all the necessary cream, the best medicine a plant has. That flower has its strength and the bee comes and gets it. It's extremely important also for the human being. You take a spoonful of honey every day and you are taking vitamins.

93 Guano, in Cuban context, refers to the palm thatch used to cover ("cobijar") the roof of a bohio.

94 Décima, a type of sung poetry (ten stanza) introduced to Cuba in early colonial times, sustained by rustic Cuban people.

95 Torolico,tocororo, (*Priotelus temnurus*) national bird of Cuba.

We have left beekeeping, my brothers and I, because the sting of that little critter is very harmful for me. They'd sting me and I'd get very swollen and therefore I don't raise them but my brother was very dedicated to it. Although he is old now—he's a very old man—but he liked producing a lot, as many as thirty beehives, with tanks flowing over. Nowadays over around that area, Eusebio [cousin] is the one who has beehives, who breeds them and takes care of them, because they [his family] are not afraid of that critter and there are wild beehives around there, so people go to the mountains and harvest them. Over at the foot of my house there was one and a man took care of it. He extracted a bucket of honey. I said, "Take care of it because I want that beehive there, but the beehive left. Honey is good for anemia. You drink parsley juice with honey and an egg yolk. You mix the honey with the rest and it stops the anemia at once. All of that is extremely important.

The jigües

There's a pool here they call the Jigüe's Pool.[96] I never saw that jigüe, but those old Indios like my grandfathers Serafin, Cancio, and all of them, said, "Careful, for that pool belongs to the jigüe;" they saw it. That's over in La Escondida, a pool they still call the Jigüe's Pool, a big pool like a river dam, where they say a little jigüe appears. And that's clear, so no one swims and no one goes fishing there.

I was still a bit young, a little kid, when they saw it the last time. I didn't see it, but as youngsters we never went there because our parents used to say, "Children, a jigüe appears around there. Bathe further downstream. Don't bathe in that pool."

The jigüe can be playful or it can hurt you. It is best not to bother it. It is best to show respect, because they're the owners of some pools, rivers, [places] wherever there is water. And water itself gives you life and it can kill you—it can drown you. The jigüe's influence is precisely that. And it has a current; it has a current.

That's all about the jigüe.

96 Jigüe: In Cuban mythology of most antiquity, these are small elves (or "little people") that are Indian spirit-guardians of lagoons and streams. Reported both to be of Indio and African appearance, the jigüe becomes a cultural mirroring of the Indian-African duality in Cuban imagery. The letters are reversed to güije in Camagüey and western provinces. Fariñas (1994) gathers the oral history of Bayamo about "jigües," whom, the tradition tells, "stunned by the Spanish Conquest, responded by concentrating themselves in Bayamo, and by setting in motion their combined supernatural power to impede the disappearance of the religion of the Cemíes."

The owl's message

The owl is a messenger. In our land, that's what the owl is: a messenger of profound darkness.[97]

This is confirmed by the sensitive ones, always. From any people. Look, also for the Abacuá, in African belief, say it is Obatalá. When an owl flies by, they say, "Look where Obatalá is going." And it's because there is a mystery with the owl.

We Indios still carry it. An owl flies over a house screeching "*chua,*" and we say, "Magnificent Mother, I believe in a Creator. Chase that creature away."

"*Solavaya!*" we say. "May that owl go alone."

That's advice from long, long ago; it's been our belief for many years.

If she flies by, say toward midnight, and she makes her sign, or if the owl comes in a dream at dawn, and she appears to you, well, that creature brings a warning to the human beings living in that place. Someone might die or anything may happen in the family. You must heed that sign and prevent and prepare yourself. Because sometimes you can petition, you can ask for more life, and the message, the petition, goes back [with the messenger]. And sometimes the decision is fixed; there's no remedy. Look, that's a system we have here. When I see an owl I say, "Magnificent messenger, I believe in life, I believe in a god. Take her, San Alejo and the great winds: let her gain distance from me."

The chipojo

There is a small animal called the chipojo. It's a long Cuban lizard.[98] It's also medicine. It has a special spirit and also transmits its current. That little animal descends from high on the tall trees exactly at noon. And precisely at that time, it eats, drinks water, rests ,and climbs back up. It goes down and up that way with absolute precision.

97 For a comprehensive treatment of the owl in Taíno culture, see also José Juan Arrom, Artes *Prehispánicas de las Antillas* (México, D.F.: Editorial Siglo Veinte,1975); and Manuel A. García Arévalo, "The Bat and the Owl: Nocturnal Images of Death," *TAINO: Pre-Columbian Art and Culture from the Caribbean,* Fatima Bercht, et. al., eds. (New York: The Monacelli Press, 1997): 120+. Arrom writes: "The owl of Caribbean folklore, a bird of ill omen that announces the proximity of death, has inhabited a mythical world very different from that of its European cousin; in ancient Greece, the owl was a symbol of wisdom."

98 Chipojo: *Anolis equestris*; knight anole.

It's a crime to use him for medicine because it requires that the little animal be killed, but his medicine does heal. It is used when you have a thorn and your hand or your arm gets swollen. You catch one of those lizards, a little chipojo, open it up, and place it on there, and it pulls the thorn out. You tie it to where the thorn is and it pulls, sucks out that thorn—similar to what copal does.

All these are Indio things, guajiro. They are our things that persist in the mountain. Some might say, "Well, why? How can that be?" But it's because they're old things and in many cases, whether it's through spirit or mere science, they're ancient methods by reason and by knowledge.

They used the warmth of the chicken, for example. When you had a very high fever they'd take a tan-colored (*jabao*) or black chicken and opened it up, remove the insides and all of that, and they'd put it to your brain or here on your forehead; they covered you in wing feathers and the fever went away. This is to say that many such things were done in the olden days. The Indios did it because, well, that was their resource.

Nowadays, as I was telling you, with science those things aren't done as much, because studying tells people different things. But those ancient methods still exist, because there are people, when their fever is through the roof and the doctor prescribes pills and it doesn't go away, who say, "Put a chicken on the forehead or on the chest," and they take that, *bam*, and do it. Yes, it's true that not everyone follows it; these are not systems people take everywhere. But neither are we who follow these ways so very few.

The tracking cure

Among our ways there is what is called "the tracking cure." With this method we get rid of worms and heal toothache and asthma, in human beings or animals. We also pick up the trail to things.[99] I had a godmother who cured the worm-infested animals and also human

99 "The tracking cure," ("Cura del Rastro") in its various adaptations, is a ceremony practiced by tracking healers or Rastreros in the eastern provinces of Cuba. In Camagüey, Herminia Padrón de Jiménez, aunt of the author, lived for many years in the Camagüey guajiro farming country. In personal correpondence (1987), she writes. "About the uses for healing "by track," an affected animal or a person, there were prayers. Different [rastreros] used a favored tree such as a guasima treee, or the guava tree. One approach [to the sacred tree] is to carve three small sticks, same size, with one end pointed and the other diagonal. The animal or human that requires the healing is made to walk with the wind toward the shade of the guasima tree. Steps in the track are observed and upon the first left foot step inside the shade of the tree, a stick is inserted, similarly for the next step right and next step left. Many claim good results." There are other adaptations of the tracking cure ceremonies.

toothaches by tracking [the illness] with her prayers. She was a person from very long ago. Her name was Justa and she was part *gara* (white) and part India. And I still have a sister-in-law who, with prayers, kills worms from long distance. I think about it myself and it's incredible, but it's true that the worms die—they leave the infested animal or human having worms, or having a toothache. She heals toothaches and gets rid of worms in animals, and in people.

If the animal is having worms, you have to tell her where—if it's on the left side or on the right—and [she cures] with prayers. I don't know what the prayers are called, because I've approached her and have asked her, "And how do those prayers go?" She says to me, "Well, that can only be taught to one person, but not two." They're reserved people with their ways. She says, "If I teach you that prayer, I can't teach it to anyone else." So they're whims but I know they have power because I've seen myself that, when she does her ceremony, the worms of the infested animal have died.

The tracking cure is a prayer in which you use the animal's tracks. You make little crosses and keep praying, and you have the animal walk and the human being, too. I'm not a *curandero* [healer] of the tracking cure but I've seen it being done. And sometimes I am able to know something through that path.

There are many things we have that are remedies for many diseases. There are prayers we say; for instance, for a dog ready to jump on you, or an ox, to a person coming to get you, [there is] "Stop ferocious animal, magnificent, ferocious animal" to an ox. These belong to the many prayers used by those Indios of former times. These are prayers they used and they were real because the animal did leave. Or you found what you had to find or had to know.

I had a brother-in-law who already died, who used to go to a house where they had a very aggressive dog and he'd say those prayers and pass his hand over the dog's head—it didn't do anything to him. I was telling you they were real, useful. Today it's no longer done with precision, but I get good results. That's the talk by prayers like that; it's spiritual. See, you can give messages like that.

I don't talk so much here about Cacique Hatuey's era, but about the more recent times up to now, because a long time has passed since the conquest. However, we continue to have prayers that are very much our own. We truly have what we've learned since the time of Hatuey and Guamá, but we also have our own teachings, because around here we Indios also made contributions.

Things belonging to the mountain are obvious. Because everything that's trees and plants and their use, comes from the past. The yuca is Indian, with its casabe and everything of hers – the moon, and the medicines, too. And we Cubans, where does our love for the Earth come from?

Tracking down a thief

Little things happen in any community and you have to respond. Nothing serious happens here, but little things like small thefts or a loss of respect.

Look, there once was a fellow who took one of my animals and I had to throw him out. It was a bird, a *guanaja* [female turkey]. We noticed that it was lost. I simply couldn't find my guanaja.

I started to investigate and I asked my altar. I'm doing my work with the altar, and just like that, all of a sudden I say to myself, "Oh, but it was this fellow who took it."

So we continued searching until we found out about his habit, until we collected the evidence. I held watch, and I saw that one night when he left to where he buried something. We went with the guys to where I had watched him and we found lots of feathers and that's how we caught him. Later it was revealed that he was a suspect in several such cases.

I called him over and said, "Look, whose is this?" and I showed him where he had thrown the feathers in a bag and I said, "Look, when you ate this, it was theft. It wasn't a gift." And I said, "And why did you take it if it wasn't yours? It's mine, it belongs to my children, my people at home."

He was one to whom I had given a piece of land, imagine, to build his little house. I said, "I gave you that piece so you would build a house and it wasn't so you would eat an animal that wasn't yours." Then I went and gathered all the people, the entire committee of our community and I said, "Well, what's your opinion?" The people said, one by one, "Well, I believe he has to leave."

So I took the case to the local authorities and they said, "Panchito, the one who gives orders there is you. In your *ranchería* [town], you do as you see fit," although the police did come and fine him for his nerve. And I said to them, "Well, he does have to pay me back."

And I took a bird—a rooster is what he had—and I took it from him. I said, "Give it to me in exchange and leave this place, and be done."

Those are the minor problems that can happen and that I need to look after. That's the only occurrence [like that] I've had. And I saw it through my altar; [I saw] how and who it was.

Mother of Water

Mother of Waters was indeed a serious thing, a power with capacity. For the old folks she was such a great thing that they held her for a saint. She was like the water spirit, and often they had a candle lit for the Mother of Water by the river on a rock. That [practice] has always remained among us. We feel that strength and I myself have ordered remedies, have cleansed people, and have told them, "Alright, look, go, bathe yourself. Give yourself a little bath with the Mother of Water. Duck into the river three times and light a candle for her and ask her to take away all the evil you might have, that you have all the spiritual water she can give you, so that the spirit this water actually is may help you to purify yourself."

Lunar planting and the humidity of the earth

For every food there is a proper time for planting. There is the waning moon to plant corn and beans. Yuca is with the waxing moon, so that's in the new moon. The boniato is in the waning; the *guineo* [banana] is in the new moon. I look for the meaning for you: corn doesn't spoil during the waning moon, but you plant it during the new moon and it spoils. When you are going to put corn into a container for the hens, for example, or for pigs and so on, in a short time, if planted during the new moon, it spoils badly. But corn planted during the waning moon doesn't; it stays hard, and the weevil doesn't eat it.[100]

100 The Taíno practice of planting and harvesting tubers and trees according to lunar phases was noted though not fully explored by early conquest chronicler Oviedo: "They always when planting do it in the early moon, because they are of the opinion that as she [the moon] starts to grow, the planted crop does the same." Gonzalo Fernández de Oviedo, *Historia General y Natural de las Indias, Ed. Bruño*, ed., book 7, chapter 1, 1991: 108+.

The yuca develops more during the waxing moon; then the plant grows bigger and the yuca tuber grows more, while during the waning moon, the yuca is shorter, fat, and without stem. We call it "stem" because it produces lots of fiber. When planted during the waxing moon, it doesn't have fiber. What happens is that, in the mountain during the waxing moon, there's more growth because there's a humidity the old ones called humidity of the earth, which during the waxing moon helps more. All the plants are with water; all the trees hold the water. And that's why in the mountain the yuca is planted in the new, waxing moon.

But that same humidity of the waxing new moon rots the corn. That's why the corn is planted during the waning moon, the land is drying and [the corn] is harvested during that same waning moon; it stays hard and doesn't spoil. So we call it "waning moon corn" because it doesn't spoil on the plant nor does it rot at home in a container. But all the corn that spoils on the plant, those you see all full of weevils, was planted during the waxing moon. That's certain. That's a system our old grandparents taught us, from whom we learned how to do things by watching and listening to them.

Wood for the house and the moon

The wood to build a house and the guano to make roofs is also influenced by the moon.[101] If it's cut during the waning moon the wood will not decay, and the same with guano. But it does decay [if cut] during the waxing moon. Bugs might not perforate the wood; weevils don't perforate it. But during the waning moon is when [the cut wood] does not decay, and available are the straight posts, the fences, the forked posts. You saw the guano over there, how it was already all in pieces. That guano was cut during the waxing moon. So I told the young man from the province, "Look, we're entering the waxing moon. Look how the guano has already fallen apart." So that's an experience you learn in the mountain, from people living in the mountain.

Corn, beans, yuca—planting conucos

One important thing were seeds—corn seed, for example. When I was a kid our grandparents kept it; they just kept it because they

101 Guano, in Cuban context, refers to the palm thatch used to cover (*cobijar*) the roof of a bohío.

didn't buy corn, so the family's seed, the whole community's seed, was a main source of life in the mountain. What is now called "self," from "sufficiency," (*auto-suficiencia*), that was our culture.

Corn was dried in the sun, there in an open canoe, and it was stored and thanks to that custom, corn wasn't scarce and it still isn't. It doesn't develop, you hear me, like in former times, because the land is somewhat exhausted and we have to add fertilizer, natural or chemical, to the corn. But the idea is the same. Sometimes the corn grows well and other times it doesn't; anyway, we're here and it goes with the traditions of the time of the pure Indio, because corn and yuca and beans and boniato, those are also from long ago. Corn and beans are planted together. You plant one row of corn and two of beans. Right now I have one planted with beans and corn . They are harvesting it now, and boniato as well. That method, which is the old conuco, is also called "racing." My father and my grandfather used to slash and burn. They cleared the land and then they set it on fire. Then they planted one row of malanga here and one of guineo there and one of yuca there, one of plantain over there—each one in its row and you went to see the field and it was complete.

Not today, because we take a field and if it's yuca, yuca it is; or if it's malanga, then malanga; if it's boniato, then boniato, and nothing else. But it wasn't done that way before. We've left that system, but I'm telling you that before we never lacked any of those good foods. That system was like a store in the mountain because there always is something. When you don't eat yuca, you can eat from the malanga and you can eat banana. and if you don't want banana, you can eat from the guineo or the boniato, because the rows were alternated, with a border of boniato around the whole. Those systems have been lost somewhat, but now we are planning to use them again. One of my daughters said to me a little while ago, "Old man, why don't you do the way we used to when I was little? Plant the field how you used to: one row of each thing and you always had everything and nothing was scarce."

That's the conuco. That's like that conuco in my dream. Wasn't that what Mother Earth told me, anyway? To maintain what we know. To dedicate ourselves to planting. That's what she wants, to scrape her back a little. She wants us to suckle from her.

That's her instruction. And I promised her that. So I'm going to pick up that system again and same with the animals. With animals the system is that each one has their own animal. You tell the child, "Listen, this goat or this cow or this ram is yours," so that people have more love toward it, because if we are four or five in a household and one alone is in charge, then the others don't care. It's better if we all look after them because you have yours and I have mine and you have to see it as your duty and I as my duty, but together.

That was a system everyone followed. The Indian system is to raise many things, harvest many foods. Among the food sources was the Puerto Rican malanga—there was the yellow malanga, the white malanga, the purple malanga; there was the *sagú* [sago palm]—all Indio things—there was the *afio* [celery] in the mountain, the *chote* [chayote], the corn named tucunú, but today called "*tamal*," and the corn. The true name of harvested corn is *tucunú* and that word is Indian—tucunú, that's the *hayaca* [tamale] we eat from the corn, tucunú, and no one calls it tucunú anymore. They call it "tamal" or "hayaca," but it's tucunú.

Corn is a brother

In corn we can also lose the variety of our seed. We must look after all that. Because there's the Argentinian corn, which is a more yellow corn, more from present days, and there is the creole corn, which is what we call the light-colored corn with the fine cob, which is our creole corn, natural Cuban corn. A little corn that grows with a very fine stem and the little kernels a bit tiny—that's the real Cuban corn. Eh, creole, (*criollo*) we call it creole corn, cradle corn. There is a purple corn and there is the Argentinian corn. And there is another yellow corn that's a bit lost: it's a yellow corn, but not Argentinian, and that one is also a normal corn from here, more yellow from former times.

Corn is an ancient brother of ours. We Indios use it in many dishes and it is our friend, our brother. Look, corn gives us around fifteen different dishes. You cut the very, very early corn and you make it into *congrí* [black beans and rice]; you make it mixed. That's two. You make it into *manencia* , [referring to thick or uplifting rich food] three. You make it into *atol* [corn beverage], and there are several ways of making atol—with or without milk, for example. Then there's the *mazamorra* [milky maize pudding], which is a corn you grate and make into thick atol that you have to cut with a spoon, with a knife. That's already five. There's corn bread, six. There's the *serence* , [mountain vernacular for pudding] which is the one you stir and let it sit to eat. That's seven. There's the tucunú, which is the tamal or the hayaca, eight. There's the *dejuculeco* of the corn, which is the corn toasted with grease and salt, nine. There's the *pinol* [roasted cornstarch], ten. There's the *fritura* [fried], eleven. There's the *chorote*, twelve, which is cooked corn like a kind of chickpea. And you make a corn stew, thirteen. You make *arepa* [pancake], fourteen. And the boiled corn, the cob which you boil— that's already fifteen and there are more, there are more, there are even more derivatives you get from corn that are extremely important. And those are the things about corn; fifteen or more things you can make. Corn is extremely important, for what we get out of it and for always being with us. We have faith in it because right now there is a shortage of rain. But corn persists. And therefore people like corn.

Oh, another one, look: roasted corn. Imagine, sixteen. That's the most delicious one. You roast it on the cob over a fire. It's the most natural and delicious one. It's easy to appreciate corn that way. You roast the corn on the cob and have it with your cup of coffee and you're eating.

For corn bread, you grate the corn and then you grate a coconut and add it. Then you cover it with guineo leaves at the bottom, you place guineo leaves at the bottom of the caldron and you fill it up, you put a few embers on top of the lid of the pot or the cauldron, and you're cooking. That's the bread. Pinol is a powder. You roast the corn and grind it. The corn arepa you cook in a leaf or paper and bury it. For the serence you use the flour from ground dry corn and you mix it with salt.

Well, corn has so many ways that we have to have faith in it. If you only have one thing to eat, what tires you the least and lasts the most is corn. Corn, yuca, and beans. It's an old element of ours. Corn is a brother.

Yuca and the sincerity of things

Yuca carries a special feeling. Yuca is an important food source. It gives casabe; it gives starch; it gives arepas.

The arepa is tasty. You take the yuca and cook it and make it dough and then into a kind of bread. Then you wrap it into leaves, add grease, and make it into an arepa. Wrapped up in leaves you bury it in the fire in a brazier and then you just leave it there and it's an arepa. It's similar to a yuca biscuit. It is very, very delicious.

Then there's casabe, which we still make when there's yuca. When yuca is in season, those yucas that are still a bit smallish are made into casabe. All that is an Indio system of good food source. Many Cuban products are made from yuca, just as from corn. You grate it and you make atol. You grate yuca and you make Cuban products. Yuca is important mainly for Indios, in the olden days much more so than now. Nowadays we have lost many things and only a few of us make casabe, since people dedicate themselves to other things. But the original [use] of the yuca, all of that is extremely important, because that's what belonged to the Indio who made plenty casabe to trade for fish with different Indio groups. Because this is how it was: there were some that hunted, the other ones grated to make the casabes, and they traded one thing for the other. It didn't cost money; it's trade, trade.

But the yuca was essential and it continues to be so. It has spirit and it has powers. You have to connect with it. In my ways I have to ask it. I've asked for my sight, as yuca's power helps the eyesight. There's a yuca plantation on my path when I go to work in my field every morning. I wash my face with yuca dew and pray, "Heal me. Give me clarity in my vision." And I wet my face with all the leaflets; I pass the little leaf over my face, leaving the leaflet right there on the plant, but I pass it wet over my eyes, my face, and I ask like you ask anyone else for good faith. That gives me more tenderness toward my agriculture and toward everything.

Our grandparents always sang those chants like the quiribá, and they used to sing in the yuca plantations, harvesting yuca or planting it, but not directly to the yuca, but to all of nature, for the pleasure of being there in the field, in nature. There was some of everything scattered throughout their songs. They sang to all of it together, not to yuca or corn directly, but to everything in a collective. The collective of the conuco of the whole crop. They sang *décimas*[102]—they made décimas out of everything, but they sang décima upon décima and corn

102 Décima is a type of sung, ten-stanza poetry introduced to Cuba in early colonial times and sustained by Cuban campesino people.

is somewhere there scattered in the décima, along the yuca and the beans and fruits; all of that they sang everywhere in the neighborhood.

The old ones sang to the food source. That's the Indian way. They made décimas to a pastry we called "arepa kittens." They also sang to the corn bun [*bollo de maiz*]. That's what the old ones of ling-ago times sang. Now I don't sing, not as much. What I do is plant, cultivate, love the earth, love the plants, because they serve us as food source and that's of great importance. You have to have love for the yuca, a sister as well.

My grandmothers—mainly Lina, my maternal grandmother, and also Manuela, my other grandmother's name—always said to me, "Panchito, ask nature to help you, to give you strength for you to work, to give you her loveliness, and for people to love you may she give you her beauty."

That's how they spoke to me and since I was little. I liked to be sincere. Everything has its respect. That chair is a chair and I call it "chair." I'm not going to give it another name. You must respect the sincerity of things. One must have that love, that necessary respect for things.

The loss of varieties

The variety of the yuca has been greatly depleted—well, also of other foods. That's a pain in the heart for me. Now there's the *sensa* yuca but the one we used to plant was the *negrona* and we called it the "real" yuca. There are still a few in some places, but not in others. I don't have any. I have sensa yuca, which is as hard; it's yuca but it grows faster than the other.

The other original one needs more time and grows less. The sensa produces more, but the negrona was the Indian yuca, the true one.

The true one is the negrona yuca. But in yuca there's also the *portoplata* yuca and yellow yuca and *yuca de marea*. Marea yuca they also called a "sour" yuca and it was only used to make casabe and to make starch—because that one wasn't used in yuca dishes; it was sour, the sour yuca. The casabe was eaten, and the poisonous starch was extracted. It's used for clothes' starch. But these yuca species have been disappearing. There are almost none left. Around some places there are a few little seeds that sprout and we say, "Look this is yuca from such and such seed."

We have to preserve all of that. That's life for the mountain.

Ajiaco is more than a dish

In any activity we do, very often we prepare the creole ajiaco, the ajiaco guajiro that's traditional from here. [103]Ajiaco is in the pot what conuco is in the field. I mean it blends everything together. Now, many people called it *caldosa*, but it is ajiaco. And everyone likes it. You add bones, you add a hen, a chunk of meat, and without exception you add creole produce. We prepare the ajiaco as a communal activity. An ajiaco is not only the food dish. When saying "ajiaco" you're saying "celebration," "family gathering," "ceremony." *Ajiaco* means getting together and mingle dancing and playing. In food there is the hen ajiaco, the turkey ajiaco— whatever you want. If you butchered a pig, you add the head and mix it all, you make an ajiaco, and everyone eats. You add malanga and if you have yam, yuca, guineo, plantain, unripe corn, chives, porro garlic, *cilantro* [coriander], parsley. Lots of vegetables are thrown into the ajiaco. It's very tasty and it's healthy for life and for the body of humanity. It's divine because it has substance; it picks up all the substance of leaves and vegetables. The broth isn't thrown away. Ajiaco picks up the substance of all the vegetables and you eat it. But you go about preparing it with different things, with herbs, with everything it needs to have. If there is some, you add meat, and that's why many people like it, because it has tremendous nutrition. And that also comes from long ago, from those ancient times. Ajiaco with casabe is very Cuban.

We have our own wines

We have our wines. For birthdays, for one of my brothers, or an uncle, or a grandfather, for instance, they'd say, "It's today," and everyone was already on their way to celebrate. Three days of celebration and dancing and eating and drinking, because those old folks made pumpkin drinks, made wine from corn.

For that wine you take the corn and put it into a clay or plastic container—not aluminum, because it releases little metal particles and

---

103 Ajiaco, is the Indian soup of Cuba. The word is attributed to a Taino construction to mean, "eye of the pepper."

that's bad. You don't put it in a glass bottle, either, so it doesn't explode. You have it buried in the ground for fifteen days and after fifteen days the lid blows off and its wine, wine for you to drink. You make it with water. You add a spoonful of sugar into a bottle of water and you fill it right up with corn—natural corn, the same one we plant—and you cover it up and you lea ve it for fifteen days in the ground in a hollow space where neither ventilation nor anything else can reach in, and when you take it out, if you're not ready, it all explodes. And the wine is made. It has to be filtered, even if you only have a little sieve or a very fine cloth. You sieve it and it's ready.

In the pumpkin you open a little hole and with the same pumpkin you add sugar and cover it. You don't add water. The pumpkin itself releases water. You add the sugar into it and you put it into a container or you also bury it in the ground and fifteen or ten days later, you take it out and you squeeze that pumpkin and wine comes out of it, guineo wine. They are called guineo wine, pumpkin wine, corn wine. Corn wine and orange wine and the one made with sugarcane are also called *chicha*.

The sugarcane one is strong. You take a little cane juice, pour it into a plastic container—also twenty, thirty days, a month—and what you get out is a drink. And that's what people drank at ancient celebrations and nowadays, too. We drank rum, but when there was no money we drank that kind of wine and everyone was happy. Good party.

The Earth is perfect. Yes, it can sink, and something can come to make it diminish. But she is perfect because she has it all. You can notice that it is us that mistreat her, as we make metals and as we use chemicals to harvest more product to make more money.

Graphic Testimony

Panchito, young cacique. Participating at the Mountain Culture Fair, Baracoa, c. 1980.

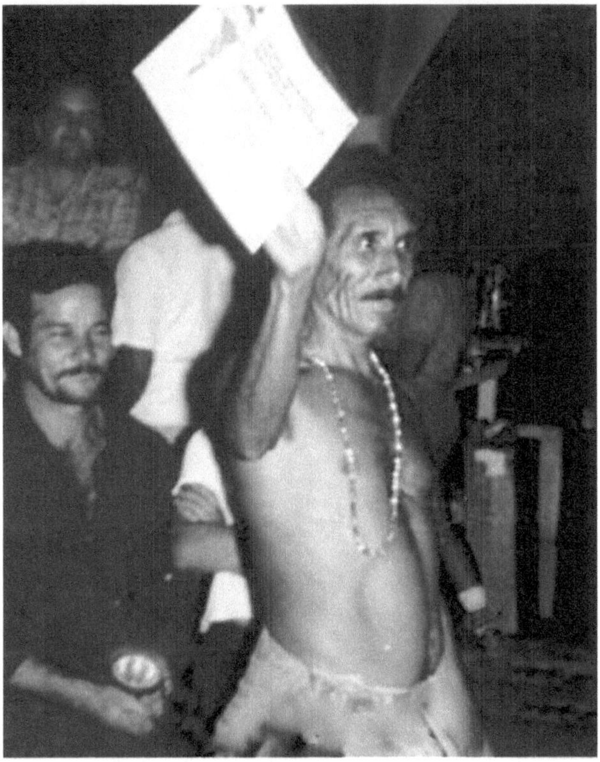

Panchito, returning from his *conuco,* first sight and image by the author, 1995. Photo still from video (Barreiro).

Antecedents of Panchito: Ladislao Ramirez Rojas, *Ladislao Cacique*, on mule, here with Dr. Manuel Rivero de la Calle, c. 1972.

(Photo courtesy of Guantanamo Historian Jose Sanchez Guerra).

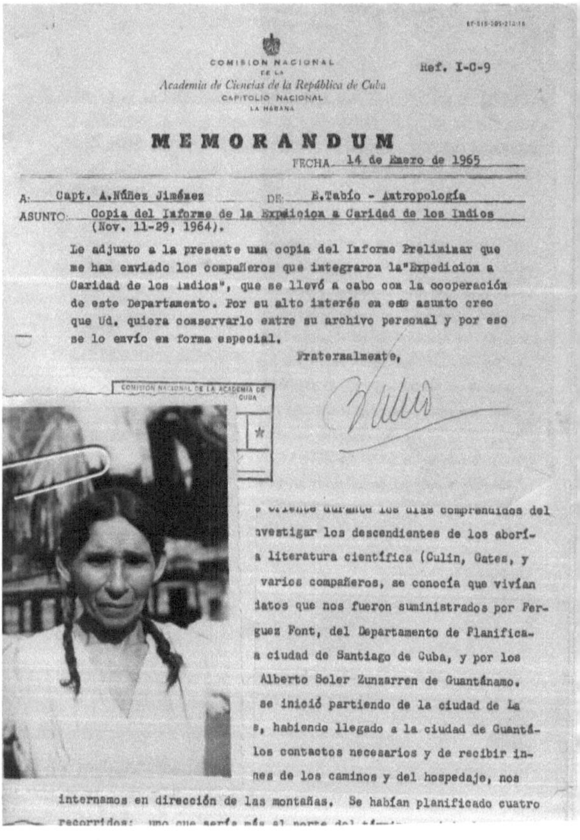

COMISION NACIONAL DE LA *Academia de Ciencias de la República de Cuba* CAPITOLIO NACIONAL LA HABANA

Ref. I-C-9

MEMORANDUM

FECHA 14 de Enero de 1965

A: Capt. A.Núñez Jiménez DE: E.Tabío - Antropología

ASUNTO: Copia del Informe de la Expedicion a Caridad de los Indios (Nov. 11-29, 1964).

Le adjunto a la presente una copia del Informe Preliminar que me han enviado los compañeros que integraron la "Expedicion a Caridad de los Indios", que se llevó a cabo con la cooperación de este Departamento. Por su alto interés en ese asunto creo que Ud. quiera conservarlo entre su archivo personal y por eso se lo envío en forma especial.

Fraternalmente,

COMISION NACIONAL DE LA ACADEMIA DE CUBA

... durante los días comprendidos del
...vestigar los descendientes de los aborí-
...a literatura científica (Culin, Gates, y
... varios compañeros, se conocía que vivían
...datos que nos fueron suministrados por Fer-
...guez Font, del Departamento de Planifica-
...a ciudad de Santiago de Cuba, y por los
... Alberto Soler Zunzarren de Guantánamo.
... se inició partiendo de la ciudad de La
...s, habiendo llegado a la ciudad de Guantá-
...los contactos necesarios y de recibir in-
...nes de los caminos y del hospedaje, nos
internamos en dirección de las montañas. Se habían planificado cuatro recorridos: ...

Memorandum of 1965 to Dr. Antonio Nuñez Jimenez, Rebel Army Captain and Director of the National Academy of Sciences, announcing the preliminary report of Dr. Manuel Rivero de la Calle about the Indian population of Caridad de los Indios

(Photo courtesy of the ANJ Foundation for Nature and Humanity.)

Anthropometric study from 1964-75, emphasizing physical measurement of the Indocuban population of Guantanamo. This study did not publish specifically on oral history.

Photo from the book by Miloslay Stingl, *"Indiáni'cerno'si a vousáci."* (Praha), 1970.

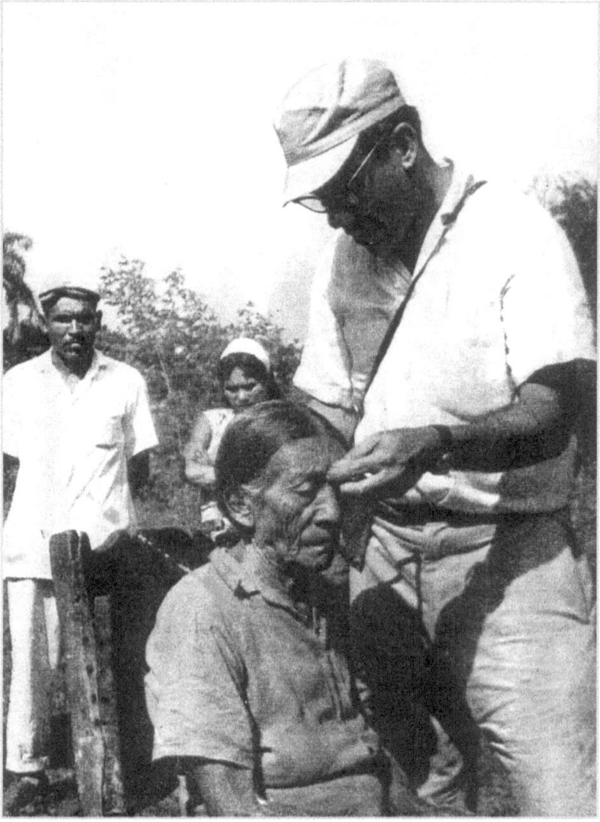

Repatriation ceremony, reinternment of Taino and other Cuban Indigenous remains removed from Cuba in 1915-16, Cemetery at Caridad de los Indios, 2003

...frecuente en el Río Toa, donde las campesinas acuden a la... ...ropas utilizando hojas de árboles como jabón. A todo lo largo ...viven numerosos descendientes de indios con ciertos hábitos primitivos.

Familia de la región de "Cuerda del Cupey", en lo más intrincado de la Sierra Maestra. Obsérvese cómo cocinan en el suelo, costumbre generalizada entre los descendientes de indios que estudiamos en la provincia de Oriente.

CON LOS ULTIMOS INDIOS DE CUBA

En los montes de Oriente.— Un poco de historia.— Las comunidades de indios.— En Bella Pluma.— Entre los indios de Yateras.— Con los indios del Yara.

Textos y Fotos de

ANTONIO NUÑEZ JIMENEZ

Bohemia 1949

...rostro que hablan de los primeros habitantes de Cuba. Son cubanos de piel cobriza, de ojos prendidos ligeramente a la manera de los asiáticos, con el pelo lacio y su cuerpo de poca estatura. Tales individuos pueden verse en lo más recóndito de Yateras, Mayarí, El Caney, Bella Pluma, Baracoa y Guantánamo. Es cierto que no conservan la totalidad de los caracteres antropológicos de sus antepasados aborígenes, porque en muchas de sus caras se ven influencias de los españoles o de las antiguas tribus africanas. Se ha discutido si tales cubanos son en realidad los herederos de aquellos indios yucatecos que trajeron los hispanos como esclavos, después de la conquista del...

Estos indios del río Yara pescan aun con sus primitivas redes, tal como lo hacían los taínos. La "foto" fué tomada en la región llamada "Cayayal".

Article, 1949, in the Cuban magazine, Bohemia, among others from Dr. Antonio Núñez Jimenez, describing the Rojas Ramirez Indian population.

Taken from the daily

JUVENTUD REBELDE

Many centuries have passed since the arrival of Asians in Our America, where they worked collectively as farmers and potters. Scholars raise the following problem: the possibility that, considering the degree of socioeconomic development even in the most remote areas of Cuba, the more or less pure genes of these aborigines will cease to exist in their descendants in the near future. Researchers from the University of Havana are working to record all these people's characteristics for the sake of the coming generations, for this task might well prove to be

an epitaph for the Amerindians of Cuba

BY RALBERTO BENDOYRO

"IN SPITE of the crossbreeding of races, the descendants of pure Indians that exist in Cuba have approximately 50 percent Amerindian genes," said Manuel Rivero de la Calle, doctor in natural sciences, on being interviewed by **Juventud Rebelde** at the Montané Museum of Anthropology of the University of Havana.

This statement by the head of the team of anthropologists doing research on the body composition and nutritional state of the descendants of our aborigines led to a series of questions as to the number of individuals or groups with Indian features and the possibility of their becoming extinct as a result of the socioeconomic changes in our country's most remote regions.

There are some 100 families of Indian descendants in Yateras, Guantánamo province. To these we may add those in Caujerí; Patana Abajo, in Maisí; in the vicinity of the Toa River, in Baracoa; and the northern slopes of the Sierra Maestra, in Bartolomé Masó municipality. It is in this eastern portion of Cuba where the team of anthropologists from the University of Havana will begin the third stage of their study in March 1982. They have already studied all the documents available, drawn up their work program and are making arrangements to coordinate their work with the Academy of Sciences, People's Power and the Party office in Bayamo.

Recently the team visited the boarding school in Las Mercedes, in the municipality of Bartolomé Masó. On their tour of the classrooms they noted that the features of some of the children from Santo Domingo, Frío de Nagua and San Lorenzo resembled those of Cuba's first inhabitants.

Dr. Rivero de la Calle interviews an Indian in Yateras.

National newspaper article reporting on the work of Dr. Rivero de la Calle. Notably, the tendency to prognosticate an inevitable extinction ("epitaph," "ultimo"). (Juventud Rebelde, 1987)

Panchito and his people, autochthonous community of La Ranchería, strength in union.

Communitarian love, celebrating the "old man."

Panchito and daughter, Idalis, cradling her mother, Reina Rojas.

Panchito, young niece, and uncle, Opublio, 2002.

Here Vladis Ramirez Rojas, son.
"I am with my father.
Our tradition will not fall.
It's for the children."

Sons of Panchito, in changüi.

Panchito with two granddaughters.

The Altar of Panchito. First the natural potencies; then San Lazaro; also, the Virgen de la Caridad del Cobre, the "Copper Virgin," matron of the island.

Ceremony of the *Macuyo* (rolled tobacco or cigar), invoking the Mother Earth, Father Sun, and the Four Celestial Directions.

Panchito in his *conuco,* 2003.
"I muddy myself in the Mother Earth.
That for me is pride itself."

Generations of persistence in the *conuco:* a son and grandson of Panchito.

Relating to Barreiro the eco-systemics of a *cutara*, the wood fiber sandal of the mountain Indio guajiro people.

Tabaco cimarrón (wild endemic tobacco), tended by Panchito and Reina (2004).

Panchito and some of his family, at the Statue of Guama, Baracoa, 2006. There is active reference to Cacique Guama, remembered in their song, "... I now leave for the mountain, with the Cacique Guama ..."

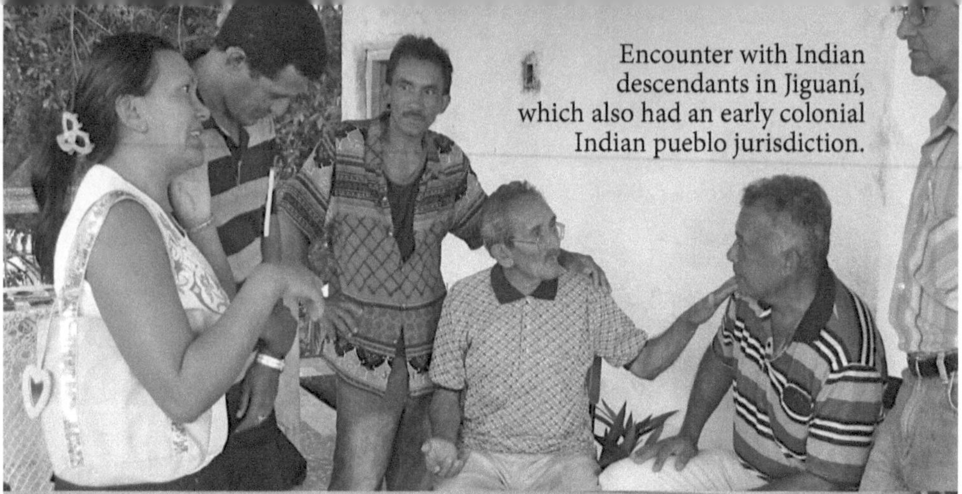

Encounter with Indian descendants in Jiguaní, which also had an early colonial Indian pueblo jurisdiction.

Panchito visiting his extended family in Camagüey Province.

Visiting family in Havana.

Extended Indian family encounter in El Caney, with rastrero healer, Pedro Rufo Mengana, 2012.

Encounter with Montoya sisters, Native family of El Caney, 2012.

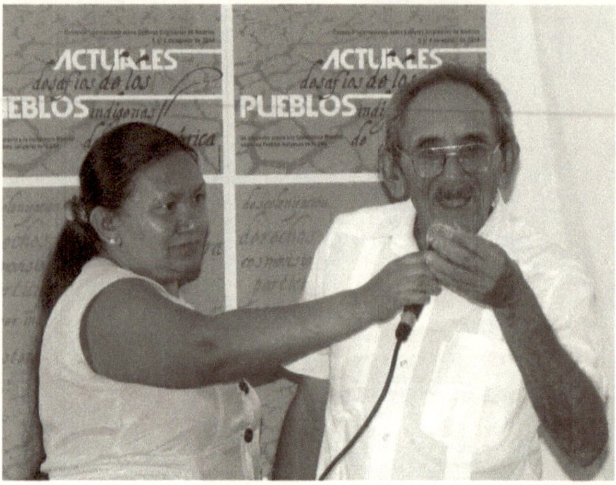

During the First International Colloquium on Indigenous Cultures, held at Casa de las Americas, in 2014. Panchito spoke on the importance of food sovereignty.

Encounter at Casa de las Americas, with Ruth Buendia, Ashaninka leader from Peruvian Amazon and Martha Sanchez, Amuzga leader from Guerrero, Mexico.

In Casa de las Americas, with Guatemalan Native representatives, Ambassador Juan Leon Alvarado, Maya Quiche, left, and Francisco Cali, Maya Kaqchikel, right.

Panchito visits the tobacco farmers of Viñales, in Pinar de Rio province, 2014. His guajiro vernacular greatly impressed the vegueros. Panchito mentions his love for the Mother Earth and they like it. “That’s the way. That’s it,” they say.

Panchito and tobacco families in Viñales celebrate tobacco. “Tobacco is everything,” Panchito says. “This man knows,” says the veguero.

Visit to Viñales Valley, Pinar del Rio Province, 2014. Panchito and Idalis locating the "heart of the field" (vega), where a ceremony should be conducted.

Encounter at the Buchillones Museum and Archeological site, Punta Alegre, Ciego de Avila. One display contains the human remains of a burial site unearthed during archeological digs. Panchito and Idalis offer a ceremony as a greeting of respect to the spirit of the human being thus exhibited.

Interview with Barreiro, Camagüey, 2013

Panchito with wife, Reina, and daughter, Idalis, reviewing the manuscript of the book, read aloud by profesora Isabel Lautin, and corrected by Panchito.

Working la Cuenta Familiar, 2009. Panchito and family begin enumeration of the many Rojas-Ramírez homesteads (*caserios*) and individual family relations.

Panchito roasting a pig, always useful.

Panchito in
contemplation.

Panchito in conversation.
Two photos.

Panchito at the Statue of Hatuey, ancestral link,
in Yara, near the City of Bayamo.

Panchito in Viñales (Pinar de Rio), with Alejandro Hartmann (right), historian of Baracoa (east) and Ricardo Alvarez (left), historian of Viñales (west).

Panchito, Reina and four daughters: Almeida, Idalis, Marta and Nazaria. Extensive families flow just from these generations.

Ramirez family, barrio Buenavista, Yateras, Guantanamo.

Family in Puriales.

Family in Tribilin.

Family in Baracoa-Jamal.

Family in Puriales.

La Rancheria-La Escondida, community relatives, 2016.

Signs – old and new – declare La Rancheria as an autochthonous community at Caridad de los Indios, evidencing a formal recognition of the Indigenous group identity, at the municipal and provincial level.

Macuyo ceremony, tobacco ritual in appreciation of the powers of nature.

In the community of La Rancheria, central to the Indian descendant population at Caridad de los Indios, ceremony is a casual and recurrent activity, mostly officiated by Panchito and other elders. Wherever the cacique travels, his folks join him in the macuyo and other healing ceremonies. Here a ceremony of community baptism, with prayers by Panchito and elder Luisa Ramírez.

Luisa Ramírez, 94, conducts a community baptism for a great-granddaughter.

Hugs and greetings end the ceremony.

Lucia Ramírez cleans the hulls using wood ash; ancient Indigenous corn processing method in use at La Ranchería.

Idalis Ramírez tries on a pair of *cutaras* (Native sandal) on her daughter, Ingrid.

Campesino discussion on planting and harvesting by the lunar phases.

Filling in the blanks of knowledge about lunar planting.

My real thing is to respect, from a child to an elder. A child talks to me and I listen.

Advice

Tradition is to do what we do

Tradition is what you have done, what I have done since childhood that I still carry today. Many say, the way you used to do things, it is not done that way anymore. But, why can't we do as we used to? Eh? Since we are human beings, we should do things well, as we should, and not be charlatans.

Tradition is to do what we do, how we plant, how we sing, eat, how we dress, how we share—all that is tradition. The tradition is natural; it is to appreciate. [It is] what I do with you, and what you do with me. That pleasure, that dream, that heat that you feel, the cold that you feel, well, I feel it, too. That you have and that I have: that is what is most beautiful. [It is] what one does right now, and in twenty years, to be still doing it, to sustain that in our cábula.

In the time of my childhood I ate even better than now, because we ate ajiaco, *calalú* ,[104]casabe, *ducunú* (corn tamal), which was Indian and became African and the Indians called "ducunú," and we ate arepa of corn, of scraped yuca, boniato, malanga.

[As for] the way to make our shoes, when it rained people wore *cutara* [sandals] made of palm frond. I have a cousin in La Escondida that makes cutara. We mostly use the cutara ceremonially these days, at dance presentations. I have a pair there on my temple. That is very Cuban, that is cubanía, although we use other forms of footwear today. We used to kill a chicken and boil it with malanga and yellow chopo, a special soup; that is a tradition almost lost, like so many things. Right here, I am rescuing many things. There are birds I will not allow to be hunted, nor the killing of *jaiba* [sweetwater crab], because some day we may have situations in our country that require that we eat like that. Imagine, if we don't prepare, what will become of us. We have to be prepared and I tell my children, "let's rescue our traditions," to prepare our land with a pick, with a *coa* (Taíno digging stick). We would open our hole to plant malanga and other tubers this way. The earth suffers less damage. When you plow the land, the soil starts to leave.

The Earth is perfect. Yes, it can sink, and something can come to make it diminish. But she is perfect because she has it all. You can notice that it is us that mistreat her, as we make metals and as we use

104 Calalú, Amerindian word, here stew made by leaves from the plant by the same name (*Amaranthus spinosus*), blending many other vegetables. http://www.enciclonet.com/articulo/calalu/

chemicals to harvest more product to make more money. Students come here and I stick them in my temple. I transmit my idea to them. "Do it this way," I say, "it will benefit and you will remember me." The land is good; it gives me food, unlike so much *charlatanería* [quackery]. Let us love the earth. Look at this plantain plant—if you knew how much banana it has given me. Panchito planted one plant but the earth produced the food through her; it gave her water and everything, I did nothing else . . . the earth gave all that to me. That is the correct way to consider it.

What needs doing is to plant food. As I consider it, the human being does not use even one-third of what the Creator gave us. Think how many fruits: guava, guanabana, mango, prune, orange, *mandarine* [tangerine], *mamee sapote*, cashew . . . meaning, we have a lot of fruits to eat. And we don't plant! Not even one-fourth. But our old people, the Indians, they did have a lot of food, because nobody ransacked a fruit tree growing freely in the forest, nobody sold it, nobody cut it down. But today, to make a peso . . . ah, a man was here that wanted to log our trees. But I caught up with him.

If rainwaters will come, all our food is guaranteed. But pay attention: what has been done in the world to this nature that gave us everything we needed for happiness—so much war, so much selfishness—has left us less than half of what nature gave us.

Children need to be taught

Children must be taught how to love the Earth; if we don't teach them, all is lost.

If we teach them only by *novelas* [soap operas] on television, that is a crime. Children need to be taken away from all that, so they can see nature, to see how plants grow. Today children see more novelas than even the adults. I don't like to watch novelas, because when you leave the novela, you emerge with thoughts of seeing strange things. To walk on the Earth, this is what we must do, this is what is good . . . .

Antiquity can guide us still, television intrudes

We displace many things of antiquity; we discard them for the things of today. But we have to grasp an example from antiquity, that those people lived better than us. Now we speak of development because you wear clothing with an adornment, you wear a pair of shoes, cook with electricity. And what do we do with all that? It's just a huge waste of resources that belong to humanity.

Antiquity can guide us still. They lived more calmly, more humanly. When the young people come to live that way, they like it. But what can anyone do about television?

The way the old people lived before us—that antiquity does not exist anymore. In that era, you could not watch television. It is not easy now to teach a young one; when you talk to them, they are watching television. There they can see it all, when a man and a woman embrace, kiss, when people kill each other, and that little one is picking up all that from somewhere else, from a television somewhere they make that, there in Havana, or Camagüey, or Mexico, who knows by whom . . . and we see it here. And what are we seeing? Will this be the calamity of men and women?

That is what I say we did not see in that antiquity. The young person, the old person would be in the good. But today you say anything and quickly they snap back offensively, because they think you mean to offend, because they see that on television. They come with an image of something else, not of our reality. That is why the things of our world are more difficult to teach.

How would we more easily control a normal way of educating?

It can't be done, a way of teaching correctly as it should be done traditionally, not sloppily, not quackery.

There are songs today, one tires of hearing those, that offend women, that offend men. Before, songs were of a more popular nature. Here in the mountain the old changüi is almost not danced. The changüi music we play is more and more for the old ladies. The young people are going to the *pum-pum-pum*, or the *reggaeton*, so that in some places the true Cuban music is hardly danced. We really can't permit that. What is Cuban is what is Cuban. Our traditions are our traditions.

The Creation of the human being

My grandfather on my father's side was right here, and there was a big avocado tree. One day we were sitting and the sun was shining. He tells me, "Grandson, and where do we human beings come from?"

I told him, "Grandfather, that would be something for you, the old people, to know."

He said, "Look, from the people before us and before all: the first man or the first woman was born from a swamp, from a little bit of mud that became a small worm, a thing that grew from what he could find in the water . . . from that was born the first man and the first woman."

This is what he was told, my grandfather said, because in the swamp many insects grow, so it would be that humans emerged the same way.

He would tell me, "The first tree came to the world, the first shrub, the first grasses—who brought these? The first grain to eat? The first fruit? It was nature itself, and then the human took advantage of it.

"And who forms both the man and the woman? The womb of the woman.

"I tell you, we are born from a drop of blood, both my father and mother. The man makes contact but that little drop of blood in the womb of the woman starts to form just like that little worm that I mentioned. It just may be true that people emerged from the lagoon. We must love the women just for that, because she is the one that actually produces the human being. There is the man and the woman, but the man can not engender. The child comes with his respiration, his eyes. The woman fabricates because that little worm is fed by her womb, her hide, her conjuncturing; through the woman comes everything."

The man and the woman come through the womb of the woman; the virgin got that moving. I would ask, "But grandfather, who made the virgin?"

"Grandson, that one came from the swamp, from a lagoon or from the edge of the sea, where she appeared. Man then appeared and they had contact, and humanity appeared. There are all types of colors, which makes us think of how much life there is and beyond life, humanity itself, because life is all, you, me, that tree. . . . [There are] so many things nature gives us, you'd think we could get along with her . . ."

My grandpa would tell me, "You hear that name of Ogun, the warrior, or Elegguá, or San Miguel, San Pedro—those are all Indians. Very few are not Indians, and they have the same color as Indians. Santa Barbara is a beautiful Indian woman, very pretty, and she likes the good. San Lázaro is Indian, too. The devil is Indian, too."

He would tell me, "The deities of the Indians have been the palm trees, the *cañambu* [bamboo], all the big trees, the mahogany, the jigüe."

My grandfather told me, "God does not exist." I would ask, "why not?" He said, "Because God is you, who brings me food. He who gifts me with anything is God, a Goddess or a virgin, or a Santa Barbara, or Santa Elena; to look upon that mountain is to see a God, is to see Martí. Thus for me, God is nature, the sun, the moon, the stars. The devil is God. Everything has to exist. Nature gave existence to all things. Because if you don't make fire, you can't eat cooked food."

Indian things have their system

I feel that we over here speak differently. We maintain the ancient Indian words more. My grandfather used to say, "We Indios are not bad. Indios are good. If anyone has been bad, it was the white people who were bad with the Indios." And it's true that they caused a lot of harm. The Spaniards slandered us; they even said to us that we were beasts, when our lifestyle as Indios has its own concept and it's a very humane system.

And what is the concept? The concept is happiness, to feel well as humans. Because Cuba's soil is so good and that is happiness. Because the elders knew that the Earth was the giving one, that she was the sustainer. They loved the Earth. And they knew her. They could say with authority, "This field over here is for yuca, this for coffee, or this for horse pasture or for cattle, this for planting corn, boniato, malanga, yam." They knew. And that's why I want to plant lots of boniato, malanga, and yuca, there on the hill, because that's my happiness. As I was telling you, it's the pride of planting that is the fire of my culture.

I tell you, I sit down on a stone to see all those plants, to watch how the wind touches the leaves, how the shining sun sets. Right now I have a tomato field that I enjoy watching; all those little plants, I count them—how many there are in a row, how many have died to replace them— and for me that's what I'm proud of, that land. You know what it means, that since my childhood I've remained here. I've never left from here and today I'm here, because, well, we're working on this testimony; otherwise I'd be in the field.

Commitment to the conuco: to plow or not?

Just this morning the conversation I had with my children was that we are now going to pick up the system of ancient times and plant in conucos. That's what my dreams tell me, what Mother Earth and my body tell me. We are going to study, to practice the Indio system of the mountains. We're going to plant a row of boniato, a row of yuca, a row of corn, of beans, like that. In the forests we make paths like we used to. You make paths to plant and you don't burn the earth. To make her produce without burning her has more strength.

Because before, in the most ancient times, agriculture had the conuco. Conuco is like a little plot, a chunk of soil somewhere. Conuco is our salvation, our mountain store, our resistance. The word *conuco* also comes from our own, from the origin. A conuco has furrows. I make a pile to sow a yam plant, I remove the soil, and then take it and put the seed in it. That's a furrow. That's what we call a big furrow. You break the soil and once it's deep you pile it up. You make piles of soil and that's the conuco. You can keep using that furrow all you want.

Now I have all of that in yam. I used it like that in February. I have a few yam plants sowed, and I dug those furrows, a whole row myself with the pickaxe. I said to the boy, "Look, I'm going to dig twenty furrows of yam there." Also, in a row down below the house, I have some guineo. Well, the conuco system is useful for people everywhere. And not only Indios, but in the mountains people already use it. People have conucos where they plant malanga; they plant yam, yuca.

And that's called "conuco," which mixes different crops in one same place. And if I'm not mistaken, it's an Indian word from the Taínos, ancestors of ours, a word originating from our people. Here everyone uses it. "What did you plant in your conuco, cousin?" someone asks. "Well, look at my *conuquito* where I'm planting boniato, planting malanga."

I do remind you that the system of conuco planting is about mixing the products, planting furrows of different products together. In the same conuco there was corn, there were beans, there was yuca, there's yam, there's malanga, boniato, plantain, guineo—all of that in the same conuco. And today you don't; today you plant a conuco, if it's yuca, it's yuca alone, and if it's malanga, it's malanga alone or almost alone.

And the thing is that with the Indian conuco there is always something to eat, because if you plant everything, you have everything. And plants help each other. You can see that there are some that go together, for example, beans and corn. Corn always grows above all the rest. When corn is ready at five months, you can gather it. But the malanga takes a year and the yam takes a year and the yuca as well. Now, boniato takes three months followed by a variety of boniato at six months. And the good thing is that little by little you're gathering food and you keep replanting the conuco. You pick that plant, that one over there you leave, and another one you sow. It's not like you're going to take everything, but rather you take that yuca plant, so you plant another one. That was the row system. That's the Indio's food store, the conuco. Right as you're pulling it out you are already planting another one. The year went by and we had food again. With several good conucos and young people willing to help, that system works and everyone eats. That's something we Indios offer to our country. It is a foundation; we offer the conuco.

But today the conuco is not like that. Now it isn't; now people harvest all over here and there and they don't replant, or they plant something else. In my way of thinking that's the reason why we lack so many things that should be easy to supply at a broad level. Because if you pull one plant but sow two, you have two plants coming further down the road. That's a system of old from ancient times, from my ancestors.

And I want to reintroduce some of those systems. Without plowing, I'm going to work the Indian conuco, plant yuca by pickaxe. Without plowing. With the coa. I'm not saying we don't plow the fields. But the coa in the mountain, the coa is good for corn, so bugs don't get it out. Take any stick and you make a coa, then you burn one end and start planting.

I don't like plowing on the hill. I've never plowed there, but I do in the flats. On the hill I don't plow because the rain comes and it takes and sweeps away the soil. Fencing in a field in the meadow with stones is one system I use. I'm now going to fence in a field, and I'm going to do it with that system so the soil doesn't wash away. So I do plow, but it hurts me having to plow the soil, break the earth. I ask myself, "What am I going to ruin the soil for?" but it must be done out of necessity. Without plowing, it looks good, though; the landscape is even nicer. You see the trees, the cattle's pasture looking nicer. You see the tall grasses moving and feeding the cattle and it's very pretty to see all that cattle eating in the pasture.

In the times of my parents no one plowed. It's like cutting the earth, cutting the wood; it's making a wound. In my parents' time they worked that soil with the pickaxe and it didn't move anywhere. It doesn't wash away when you don't break the earth. The soil washes away when it's softened and it leaves. The rain comes and *wush*! But if the soil is hard it stays put. And the plant grows just as well or even more, yes, because it has strength. The conuco that feeds from plants or animal fertilizer is better. The unbroken earth is strong. Look, if you don't fertilize plowed soil, it doesn't yield, but firm soil does. The conuco keeps giving and giving. You can count on it.

Pay attention to the children

My real thing is to respect, from a child to an elder. A child talks to me and I listen. The other day I was on the road and a boy says to me—he was a little preschool boy—"Panchito," because everyone calls me "Panchito"—"what are you going to do?" I say, "I'm going to weed tomatoes." And he says, "And with this heat, you're going to weed?" "Yes," I say. And he asks, "And the plants don't get sick with the sun so hot?"

And that's when I told myself, "This little kid is indicating a good thing. It's true that with this hot sun plants can get sick."

So I said, "How intelligent you are!" And I put my hand on his head like this and said, "You are intelligent. Keep growing so you can help me, because I'm already old."

I love youth, childhood. I like seeing those children running, I enjoy watching the kids. I have a young grandson who already plays maraca and dances to the tres, with us playing for him. For me that's happiness, that's a tremendous culture we have here of singing together and knowing ourselves.

I would say to Fidel

I would say to Fidel that if there is a communist man, I am one, because if I butcher a pig everyone has something to eat, if I butcher a goat everyone has something to eat, if I have produce I give it away. Some tell me, "But old man, don't give it away." But I say, "To my own community I give, I don't sell it." I don't know what it is to sell a pound of beans. You go and say, "Panchito, I need a cup of beans." I say to you, "There it is, take, take it with you." "How much is it?" "No, no, take it."

The community the way the old people ran it, and actually the way we still continue to do more or less, is true communism. It's a community of humanitarians. I would like that Fidel found that out, that Fidel found out.

I am humanitarian, humanitarian, humanitarian, and as I told you, I love the land, I love children, animals, plants, water and everything there is on Earth. Therefore I understand what Fidel wants to do with Cuba. And it's not easy; it's not easy. Look how many years he's been struggling and with so many against it.

If I'm useful for something to my country, I'd always tell him, "Well, it'll be done." Because from around here you'll never see a traitorous Indio. I'd say to Fidel that we're still present. And if we are useful to care for and defend this mountain, this Cuban nature, we're here, always with our Cuba.

Sometimes I see youngsters wasting water and I say, "No, that water is not to be wasted. What you must do with that water is give it to a plant over there, look, to give it strength, to give it freshness so it develops; let's throw it over there. No, don't waste it here." Or I say, "It's needed to bathe an animal, for a horse, a mare, or you. Bathe yourselves. Drink it yourselves."

Because this Cuban water wasn't brought to us by Spain. Nor by Americans. Nor did a Russian bring it. I ask myself and I have asked many intellectuals, "Tell me, where did water come from, where did wind come from, who brought the wind, who brought the sun, where does so much clean water come from, who made it, who was the man that made it?" And no one can answer that question for me.

Everything man makes passes, but what nature makes doesn't. The sun doesn't go away from here, the earth doesn't go away, water doesn't go away, wind doesn't leave from where it's born. Those powers do their duty by giving to us all, big, small, powerful, rich, humble, poor, white, black, Indio, Asian—giving to the whole world. And I carry that in my mind. That's why I take care of all that; that's why I care for it so there may be peace in our community, so they don't fight or harm each other in any way. So that my generations may know what a mountain is. And so we remain united.

Therefore, to the leaders of my country, [I say,] "Here we are planted." That's how we are and how I am. And I can say that in my ways I'm an Indio from my community. And I'm from right here, from the mountain, from my Indio culture of Cuba, in what is called La Ranchería.

I believe in patriotism

I believe in patriotism, in that of Maceo and Martí. In spiritual sincerity, in humanity, we have to feel for those people who fought and couldn't live to see what they fought for, because we'd like Martí to come back to see him in person, but the truth is we all carry their spirit. I feel it when I think about that criminal blockade the United States has imposed on us. It strengthens us. I say like Martí, who didn't surrender to the Yankees nor to the Spaniards, who chained him up and later killed him, and he never surrendered, that for all the blockade there might be, we do not surrender at anyone's feet.

We do receive the friendship of those who offer it, but we are proud. As Camilo used to say in a phrase, "Cuba is tough."

That means that Cuba is respected, that Cuba isn't afraid of anything. Those are very beautiful strong words, and why our historians can write about brave men. Because we in this country know what we do; we know what's wrong and we know what's right. And to the bad one we say, "You're bad." And to the good one we say, "You're good."

And I say those men who fought in the war are dead now, but in spirit they exist. Because Maceo, Camilo, and Martí, and so many others who died, could they have died because they wanted? No, they died so we may have freedom, because I have my way of thinking and I have my culture and my country respects it, and that freedom we have to maintain.

I would like when I die

I would like when I die that they bury me on my farm because our minister, the beloved Fidel, said that the farmer who would like to be buried on his farm could do so. And I would ask the very state, if it allows me, that I might be buried in a little spot, in a little corner that's mine. For me the greatest joy, which would be sadness in my children's eyes, would be my grave under a tree. It would be a jagüey or a cupey. To rest in what's mine and in spiritual existence help my family from there so no one comes to bother them, so all who want to bring me flowers may go there and plant flowers for me, plant trees and plant edible plants and plant anything. It would be like this, "I'm going to plant this guineo plant at the foot of my father's grave."

It might be painful because I say that to my granddaughter and she says, "No, no, because you are here and to see you there in your grave and not see you alive, that's painful."

But I say "no, no, no." May it be a delight for the family to have me buried there. It wouldn't all become a graveyard because, for instance, it's prohibited. But if the state, respecting a person who has never left that little piece of land, [said I] could be buried there, that for me would be the most significant in my will.

Advice and memories

At my age now, I'd tell my children to follow my example, to get along with everyone, to serve the old, the young, the earth, the trees, to have faith in the Cuba we have. If I happen to disappear one day, may they respect our Cuban government and may they respect humanity.

I'd tell them that if a human being comes to their door, asking for a bite to eat, not to deny it and to share what they have—the little they have, always to share it with other human beings. That's what teaching one's children means to me. May they know how to learn those teachings from their parents; may the children know how to learn them, the grandchildren, the great-great-grandchildren. Not to be selfish and not to envy anything from anyone, but to appreciate what they have, to have love for mom, for dad, for a sibling, love for the one who helps another person, who helps a doctor, a teacher. May they know that here everyone who studies is a teacher, a Martí.

I say it like that to a teacher in Caridad, "You are a Martiano, a teacher is a Martiano." Which means you can't mistreat a child. A teacher who mistreats a child is not Martiano. A Martiano can't do that. He's a good teacher, therefore he is a Martiano.

That's what I want with all the children I have, that they be humanitarian, that they be happy. The humanitarian suffers but is also happier. He knows how to love. That's why he knows more about true happiness.

If they are going to remember me, well, may they remember that Panchito was a fighter—against bad things, because I was against the dictatorship, against the abusers. Panchito was a fighter and Panchito wants his children to be fighters, to live from what they do, not from anyone else's sweat, to respect another person's property.

Well, I would tell them to respect love. Love is complex and it belongs to each and every one and it's capricious. Love belongs to the whole world. Love is a dream. The truth is there should be no one getting between a man's and a woman's love; that should be respected.

Sometimes there are family situations. The best is not to get involved in the problems of families, as long as they follow the tradition of love and respect. Children belong to their parents and grandparents. They must be cared for.

Healthy and natural is good. Moral behavior has always been respected in love. There are things between man and woman that are only theirs to take care of. Pleasure in love has to be respected. Every human being has a principle to love and we all do it, is dedicated to the love of a man and of a woman, with the respect of the people from the past. That's the most beautiful thing of humanity, getting along and treating a person with love, and giving advice that way. Giving advice to a person is love. Everything has its love and that's what I have.

That's what I want from my children, that they respect everyone, young and old, That they love their mountain culture, which gives us value. And that they respect the most ancient there is, Mother Earth and Mother of Waters and the jigües, and that they respect the trees, because each one is family and each one is a being, and that they respect all of nature that loves us so much.

To end, I mistreat no one and I want my children not to mistreat anyone. That's the way my grandparents raised me and that's what

I ask from my children. We are going to continue. As Indios and as Cubans, may they follow the example.

Today I say this with complete faith, with all the honesty of my heart, that all persons, old, children, all are equal to me. What runs inside of us is blood and it runs in you, it runs in all of us who are here, and we can all feel a blow.

To my scientist compatriots

I'm a farmer, but scientists I've met several, a good number who have come here. To these among my compatriots who have been able to study and know how to analyze things, to them I'd also say, "Companions of the intellect, let us resist, because one day, we'll triumph." I would remind them all that so far we haven't starved to death. And we are not going to die, because we have intelligence and knowledge to produce. Because here the guajiro farmer of the mountain, we do with our crops what no one else does. I'm telling you that our government and our scientists ought to be confident that their farmers are not going to disappoint them, nor are they going to let them down. We are going to keep ahead, no matter what the cost and we'll be helping our country with corn, beans, raising animals, with trees, with anything there is to defend it. Because we haven't gone to bed without eating yet, which is the best thing we have.

So to my people, to the scientists, let's tell them that we are moving forward, forward, that they be with the people, because the longer in the struggle, the more experience we gain of what Cuba's enemies do to us and also of what nature does to us, the good and the bad.

Now, with certainty, drought is punishing us in our region. It's a natural problem, but it's beyond normal. We see a change; we see that there's a change in the weather. There's more heat, less water. This is a drought that's punishing us, although until now there has always been a little rain that allows us to eat. I would tell the scientists not to faint, as there always is a bit of rain that helps our little plants. But you can see problems in nature. I would also tell the scientists, "We have to gather a lot of experience from nature." That's a message I would tell all technicians, too, all engineers, all scientists. We must tell them we should work with nature, with the weather.

To the scientists, listen to me—hopefully they'll hear me—that we work very close to the Earth, with much tenderness toward nature. We're in stranger times, it seems. Like this drought that is hitting the eastern part, not only in the Guantánamo province, but the provinces of Holguín and Granma and Camagüey: my people who live there tell me that they are very affected by the drought and thousands and thousands, millions of human beings are suffering. This drought is terrible, beyond normal, and hopefully the scientists will help us. We have to know, we have to find out what is happening. Why are the changes happening that we are seeing but had never been seen by this farmer's eyes before?

From nature and from the potency of humans, much is to be learned. I make this call to them, scientists, engineers, that they get hard to work with nature. Because they are intelligent, they are intellectual, they know. I send this message to all of them—Cubans and from other places—to encourage them, give them more humanitarian spirit so they discover what's happening with nature. I believe that's a sincere message Panchito sends, who is already old and feels sick and with bad eyesight, but is friend of all of humanity.

I want a wooden cross

Here there's a system where everyone who dies gets an iron cross. But I like it that the crosses of all my relatives who have died were made out of natural wood—not from the iron that was made by man, but from the tree made by nature. Man can plant trees, but nature makes them. And I ask my children that when I die they place a cross of wood for me, not even made of wood, but from sticks, which is the most natural. I believe I'm within my rights as far as humanity and nature goes.

I support science one hundred percent; they have discovered so many things and scientists continue discovering. They have incredible ideals, thoughts, dreams. I also believe a writer dreams what he's going to do. You dream and you do. But what man does never surpasses what nature does, because man is also natural and nature is more powerful. Man counteracts nature, deforms what nature does, uses what nature gives him, but nature is more powerful. And the mountain is sacred—the mountain, the mountain is very sacred for all the medicine it gives us, with the will to heal our bodies. With the nutrients it gives to all types of animals, it gives to the insects and it even gives to the land, because of each leaf that falls, rots, and gives strength to the earth, and we are following this system. We take the leaves of many trees and spread them on the soil so it produces more. That way we help the Earth.

And for you, what I have

My bohío is over there and I wish all of you could be there eating with me and saying, "Panchito, how do we do this? Let us plant. Let us make medicine." That's what I have for you, my brother, everything that's mine.

I prefer the humane; [I prefer] directness. That's why for me you are a son, a nephew, a brother. On other occasions, when they have come to interview me like this, I got a headache. The questions are difficult, strange. But with you, they're good, they get to the point. So if some day someone asks me, "What was Barreiro like?" I'd say, "A brother who wanted to know, but not for the sake of knowing, to help. A brother who wanted to ask questions, but wanted to know for the good, and well, I submitted." Not easily does one yield to a long interview like this. They're very personal things. But with Barreiro I've seen something different. These talks are very good for the human being. We are both awakening, waking each other up. The words you help me gather, presenting them this way, that's medicine. If someone takes us seriously, they do something for Cuba, for the Earth or for a child, for a grandmother. In this we also make medicine. Because spiritual medicine is to do the good. Nature gives it to us; the powers give it to us, to help people. And the songs, and the sweet and tender word with wisdom, can heal; they make medicine.

Sometimes I tire of speaking

At times I grow tired of speaking. I think I'm boring because we've already been talking for quite a few hours. And we've talked like this with the machine and without machine at other times. Well, for me it would be a delight to keep talking about life: what you know and what I know, to reach a good understanding. That's a long experience and you learn and learn.

I know that I'm valued by intellectuals because of what I know of the natural, the natural in life. But for some people I'm worth nothing. There are those with jealousy and envy. But oh well, hopefully our leaders and our young people will listen to my recordings, to see if I am like what those people say. Because I have great understanding of humanity, of what beauty is, what nature is, what the Earth is. I'm like that poem that emerged unexpectedly. Well, to each one their own adventure, to each one their own joy. Each star has a name in the sky and each human has a name on Earth. Each human is born with a destiny and sometimes our destiny is dreadful, but sometimes not. Sometimes there are people who make our lives miserable and others who love us, embrace us, and all of the intelligence doesn't belong to just one person. Intelligence has many parts to it and it is not given in order to do evil things but to do the good.

I say thus, even though I have had little education. But with good parents like I had, parents watching the Earth their entire lives—the crops, the clouds, how the rain is doing—that's why Panchito knows so many things. How come Panchito knows how to baptize, knows how to give a mass, knows to say a prayer, knows the Indio beat of the grandparents and knows how to make a bembé, knows how to make medicine, to work the land, to plow, to handle an ox, knows how to clear the land, knows how to do so many things that I'd get tired, I'd spend three days long, twenty days saying all I know? Why? Because my grandparents and all my elders taught me something.

They would say, "The Indio makes things. He doesn't always run around buying and buying." And nowadays that advice is still worth it; we must know how to live from the Earth, for the blockade or the fight with the Americans, as well as for how nature batters us. "You must prepare yourselves," I tell my children, and the cultivation, the system from the past also has its science. And that's why I know how to live my life and my grandchildren's. I have a grandson who says, "Old man, Grandpa, tell me your story. I like listening to you talking. I like listening to you explaining to the young."

I tell young people, "Look, when you arrive at a dance, where you might have a loved one, where you're in love, try to concentrate on the necessary respect with your girlfriend, with your beloved and with your own people."

It's better to attract people, to be an example, I tell them, than to seek, to push. And I tell them that love is simple and love is profound—simple and profound. When you fall in love with something, they kill you for it and you're still in love. There are so many beautiful things life has and I'd tell you today—here talking, you and me—he who understands life knows life is a tragedy. It's sad. We enjoy ourselves and we can be happy, but we lose people and what remains is the example, the spirit, and what you taught a young person.

Who understands life? There are few, but I can say now there is someone who understands me. You were just saying, right from your heart, "You are a natural Martí," and asking if I allowed you to call me that. But brother, calling me "Martí" can't offend me; calling me "Martí" is making me grow, like me reaching the sky, arriving in space, to the moon (I don't say to the sun because I get scorched); it makes me grow. You understand me, who we are and who I am. And Martí touches us. Martí was a spirit of ours, a sun of Cuba. And with so much spirit of ours! And, without offending Martí's spirit, I'd answer, that you calling me "like a Martí" is a great thing for me.

The greatest thing I carry today in my heart is you telling me that for you I'm a like a Martí. That's one of the greatest words I have received in my heart and in my poem. Martí was so much. I'm one. One Panchito Ramírez. That's all. Certainly, I'm sincere, I like children as Martí liked them, as Camilo liked them, and I love everything. Everything there is on Earth I love.

Finishing with Martí and the Earth

I think now we have indeed finished this work. Martí is loved and the greatest Cuba has. I ask Martí and I love Martí because Martí stretches over the Cubans' sky. Martí stretches over the four Cuban directions and from the earth to the sky.

And the Earth. I ask the Earth as well. Because beyond what humans do, there she is. I've asked the Earth so much. And the Earth tells me so many things.

That woman I told you about, the brown one who came to me in my dream, was Mother Earth. She has her voice. She has her heart and she likes to be loved. She's a woman; she's a true woman. That's how I saw her. In one part of that dream she had a black machine, big as this house, a sewing machine for clothes and I watched her like that, sewing for a long time in silence, and all of a sudden she stops working and says, "Sing a song for me." I said, "No, I don't know what song." She says, "You do know." Then, in the dream she gave me a letter and because of her letter I came up with a song in that very dream. The song goes like this:

> "The letter you sent me / I keep it in my heart / I keep it as a memory / that you were my life / that you were my love / that you were my love. You sent me the news / that your bed had broken / where only I could sleep / on your pillow remained the memory/ where I was memory / where I am memory / that you were my love / that you were my life / that you were my love / that you were my life / that you were my love . . ."

Then Mother Earth inside my dream came and embraced me and I asked, "Are you pleased?"

"Thank you for your love," she said.

And since she knew I had listened to her and I had understood her, she remained there satisfied. And that's where the dream stayed.

She knew what she wanted me to do, and not only I, but many people, all Indios, all people. And I did indeed promise. I promised her.

[Silence. Panchito remains silent. He no longer speaks.]

Barreiro: "Are we done?"

Panchito: "Done.""The American intelligence is

Epilogue

February 2001: For the first time in forty years, Panchito travels to Havana, invited to initiate the annual International Festival of Cuban Tobacco (Habano). In Havana's Cathedral Plaza, filled with the most prominent international aficionados to the fine Cuban leaf, he offers a prayer, his prayer of tobacco.

Panchito lights his cigar and invokes his traditional tobacco ceremony, blowing tobacco smoke to the four directions. Accompanied by twenty-two of his people, all residents of La Caridad de los Indios, he sings his prayer to the Seven Powers, essential expression of his love for the cosmos through the Four Sacred Directions, our Mother Earth, Father Sun, Grandmother Moon, the Water, the Winds, and the Stars. Beautiful women's voices accompany the community's prayer songs.

The duality inherent in tobacco is not lost on everyone; there is Panchito's reminder to privilege the Sacred over the Profane.

Five hundred years after contact in the Americas, it can be said that the world still witnessed the persistent Taíno idea that the simple, profane pleasure to many, tobacco, in its true mandate is a sacrament of our cultural inheritance, central, as Panchito has indicated, to all spiritual communication with the natural world.

Travels with Panchito: Bariay, 2009

Vadlis Ramírez, Panchito's son, teaches the young actors the proper ceremonial smoking of tobacco.

Authenticity becomes a question as the stereotyped image of "*Taíno*" can obscure reality. Erudites mired in 1492 like to ask, "But, is Panchito *genuinely* Indian?" Panchito ignores such inquiries, asking questions of his own.

In Matanzas one time, we passed a show portending to depict "Taíno culture." It both bemused and irritated Panchito. He identifies with the term "Taíno," though he is not possessive of it, having come to it later in life. But in his pride and certainty that his elders represent the Cuban Indian people that first met Columbus, scantily clad young women and whoop-whooping young men did not sit well with him.

Years later, at a tourist-oriented coastal park, this time in Bariay, in the eastern province of Holguin, the outdoor theatre featured the theme "Encounter of Two Worlds." I sat next to Panchito for the formal performance that included a "ceremony" consisting of scantily clad "Taíno" young women prancing around while the feathered and painted young "warriors" yelled and whoop-whooped, palm to lip, in the Hollywood movie fashion. Panchito got fidgety. "*No puedo con esto*," he finally whispered to me. "I can't go with this."

Suddenly, he was on his feet and walking into the choreographed pageant. "It's not like that," he said to the youngsters. "*Lo Indio nuestro no es así*." He motioned his son, Vladis, for a *macuyo*, or rolled cigar. "Young people," Panchito said. "Let me show you how we burn tobacco, how it is actually done, then I'll show you our Indian two-step."

The photos capture the moment. The young group of performers, several actually from local Bariay Indian families, reacted well. They stopped their show and invited Panchito to instruct them. After the macuyo ceremony, they also requested more formal discussion and instruction from the cacique and his main helpers, Vladis and Idalis.

A teaching takes place, from the stereotype to the reality of Indigenous Cuban practice.

The old Indian dance step, still remembered at La Rancheria, is also shown to the actors.

Appendix

Foreword from the Spanish Edition

The Book of Panchito: A Cuban Indian Testimony to Splinter Oblivion

> found in an Indigenous headdress.
> How is it not seen that the same blow that
> paralyzed the Indian paralyzed America?
> That America will only walk well when
> the Indian walks.
>
> —José Martí

I

Recently, a small children's book written by a collective of young Cubans came into my hands. Through an illustrated history, the authors introduce the mythical world of the Native Antilleans to young readers. They rely on the fictional narration of a Cuban countryman named Panchito.[105] The storyteller is inspired by Francisco Ramírez Rojas, an undeniable Cuban, more than eighty years old, a Native of La Ranchería, a remote village deep in the Guantánamo mountains, at the eastern end of the island.

We already knew of Panchito through the publication in 2001 of the first edition of *Panchito, Mountain Cacique: Guajiro-Taíno Testimony of Francisco Ramírez Rojas* —a necessary book—whose second edition I have here the pleasure to introduce.[106] The author, José Barreiro, collects for us thoughts, ideas, and memories of a man who is a descendant of the first human beings who inhabited our archipelago—the Arawaks—who already knew *Cuba* as the name of the land they inhabited, and who constitute the original seed for our identity, a reality mostly and regretfully dismissed.

---

105 Meibol L. Diez Sánchez, Carlos Díaz Caballero, and Karel Ducasse Manzano, *Panchito's tales: The seas and the fish* (Havana: Gente Nueva Editorial, 2014).

106 José Barreiro, *Panchito, Mountain Cacique: A Guajiro-Taíno Testimony from Francisco Ramírez Rojas* (Santiago de Cuba: Catedral Ediciones, 2001).

I admit that when I was called to preface this book and after meditating about the enormous responsibility I was assuming, many ideas, memories, and anecdotes from my own experiences at different moments of my anthropological activity began to surface. I started in this work back in 1966, guided by my maestro, Manuel Rivero de la Calle. Thanks to him I was able to actually meet native Cubans whose Indigenous origins are almost ignored, even denied.

It is known and accepted that Cuban culture results from a gradual mixture that developed through diverse transculturation processes during the complex history of the Cuban archipelago.[107] Within this complex reality, however, our bond as a people with American Native lineages has been particularly denied or severely cut. This happens because of the belief, very rooted in most Cubans, regardless their cultural, educational, or scientific level, that the original population of the archipelago was completely exterminated during the conquest and the colonization processes. From that criterion, our contemporary biology and culture would have little or nothing to do with our native ancestors, seemingly a "mysterious" people "lost in time," whom sixteenth-century European science classified as subhuman, backward, brutish, inferior beings of many vices, lazy, incapable of order and progress.

II

Much has been written on the false idea of a biological and cultural extermination of the first inhabitants of Cuba, but the results of investigations rejecting this unjust misunderstanding are hard to access. Most regularly, the mass media, schools, and exponents during scientific and popular events continue to repeat that we are the product of a biocultural mixture of European and African—an affirmation that excludes other components.

There were and there are those who affirm that nothing related to our Indigenous ancestors exists in the contemporary Cuban people. This forces my meditation on the responsibility I assume to express my

107 Martínez Fuentes: In this work I take up some ideas already exposed in the article I wrote with Julia Leigh Radomski, "The original people of Cuba: A legacy forgotten or ignored?" ("*El pueblo originario de Cuba: ¿Un legado olvidado o ignorado?*"), which was published in the Cuban magazine *Espacio Laical*, No. 3, 2013.

ideas on the subject, because doing so forces me to dispute this widely disseminated false "truth" of extinction, which evidences nothing more than violence and the intent to use official history to erase reality.

In the book *History of Cuba, 1492–1898*, published in 2008, historians Eduardo Torres-Cuevas and Oscar Loyola Vega close the chapter dedicated to the aboriginal communities of Cuba with a lapidary paragraph:

> "The studies on all the centuries of the human being in Cuba are still extremely insufficient given the limited sources: the conquerors' testimonies and the archaeological remains. Their living world hardly reaches us, through the reflection of fainted rays of light; a faint reflection of a dead culture that can no longer transmit its intensity."[108]

Given that optic, it would seem not even possible, for its lack of practical reality, to study that inheritance. As Cuban linguist Sergio Valdés Bernal has stated: "Scientists in our country annihilated the aboriginal Antillean world quicker than did the colonists."[109]

Undoubtedly, Cuba is Cuba because of all of its integrated components, biological and cultural. It is an amalgam of the multiple mixtures, first among the aboriginal groups that converged from north, center, and south of America; then between Indigenous and European people; between Indigenous and African people; and even between Indigenous and Chinese people. This has been the dynamic of our human population in its continuous creation. This is why our Antilles came to be referenced by Alejo Carpentier as, "the womb of America."

The first population census in Cuba, conducted in 1774, did not register any Cuban Indians. And we know, as has been pointed out, that Alejandro de Humboldt made no contact with any Indians during his investigations in our lands, which of course does not prove a lack of native existence. Today we know that many native Cubans survived the genocide, the majority taking refuge in intricate and inaccessible regions, while others simply continued living, incorporating new customs and adapting their behavior to the new times. This adaptation

---

108 Eduardo Torres-Cuevas, Oscar Loyola Vega, *Historia de Cuba* 1492–1898: *Formación y liberación de la nación*, 2nd ed. (La Habana: Edt. Félix Varela, 2008).

109 Sergio Valdés Bernal, *Inmigración y lengua nacional.*(La Habana: Editorial Academia: 1994).

does not imply that they forgot their own ways or stopped being themselves.

The cruel and bloody conquest and colonization processes that started at the end of the first decade of the sixteenth century cannot be denied. Imposed heartlessly, these greatly impacted the aboriginal people, who were divided by the system of *encomiendas* (human-land grants) and other distributions, and were forced to work in residencies and in the rivers washing gold, all the while being "evangelized."

Diverse estimates exist for the population inhabiting Cuba at the time of occupation by Diego Velázquez and his soldiers, but it is clear they became the first *cimarrónes* in Cuba. García Molina writes:

> "...from that big mass—most of it—that which was not repressed into slavery, one can barely have news. It is only known they escaped, in more or less numerous groups, deep into the high and thick mountains where were very difficult for the Spaniards to access; also, that they hid in the coastal swamps, in the small cays, and in other remote and abrupt regions. In those conditions they lived throughout the whole archipelago, far from any Spanish establishments—a few, for many years—without having news for a lot of time, about the abolition of Indigenous slavery since the middle of the sixteenth century. Many escaped and they even united in *palenques* together with African slaves."[110]

Adds historian Hortensia Pichardo:

> "For a long time, it was said that the aboriginal people of Cuba had accepted the conquest gently and that their form of rebelliousness was to let themselves be killed or to carry out collective suicides. Nothing is further from the truth. Since Hatuey was burned at the stake for opposing to the Spanish invasion of Cuba, the history of the island is full assertions of Indigenous rebellion. One of them was that of the famous Indian cacique, *Guamá*, whose uprising and death are mentioned by several documents of the conquest period."[111]

---

110 José Antonio García Molina, *El indígena en la identidad cultural cubana. [The Indigenous in the Cuban cultural identity]*, last modified September 16, 2012, http://librinsula.bnjm.cu/secciones/278/nombrar/278_nombrar_4.html.

111 Hortensia Pichardo, *Documentos para la historia de Cuba.*, vol. 1 (La Habana: Editorial de Ciencias Sociales, 1973).

Numerous citations are to be found in the documents, but just as the Spanish colonists out of convenience preferred to remain silent in the face of reality, many investigators have preferred to reinforce the voice of the colonists and the elites, instead of confronting actual texts and their evidence.

Regardless, in recent times, anthropological studies have also contributed to the perspective that denies "the myth of the extinction." Various efforts, carried out by Cuban and foreign investigators, have offered valuable metric, morphological, genetic, and cultural data derived from studies conducted from the nineteenth to the present century, mostly in the eastern region.

Rivero de la Calle, who dedicated a large part of his scientific work to the study of aboriginal descendants in Cuba, highlighted that these populations were reported for the first time in 1847 by the Spanish geographer Miguel Rodríguez Ferrer, and decades later, in 1875, by the German ethnologist Adolph Bastian, who conducted studies and observations in settlements in the eastern region of Cuba, such as Baracoa, Maisí, Yateras, El Caney, and Santiago de Cuba.[112]

As well, Juan Luís Epifanio Montané y Dardé, who is considered the father of Cuban anthropology, visited the region, interested in knowing about the presence of the Indigenous elements after four centuries. Montané made contact with the Indian descendant José Almenares Argüello, who professed to be 108 years old and informed Montané that his predecessors were also from the the same place. Montané published the results of his anthropological study in the article "Survival in Cuba of the pre-Columbian Indian Ethnic Type."[113]

Additionally, the American investigator Stewart Culin (University of Pennsylvania) organized an expedition to Baracoa, with the goal of finding "wild Indians." In his 1902 article "The Indians of Cuba" he reviews the Indigenous somatoscopic features presented in the population residing in Baracoa, their traditions, and the pre-Columbian tools they possessed as legacy.[114]

---

112 Manuel Rivero de la Calle, Antonio J. Martínez, Oscar Tejedor, Héctor Soto, *Desarrollo, situación actual y perspectivas de la antropología física en Cuba*, Estudios de Antropología Biológica IX: 83–105. For "myth of extinction," see José Barreiro, "Indians in Cuba," *Cultural Survival Quarterly*, Cambridge, Mass., 1989.

113 Luis Montané Dardé, "Supervivencia en Cuba del tipo étnico indio precolombino," *Anales de la Academia de Ciencias Médicas*, Físicas y Naturales 47 (La Habana: Imp. Militar, November 1910): 714.

114 Stewart Culin, "The Indians of Cuba," *Bulletin of the Free Museum of Science and Art of the University of Pennsylvania* 4, vol. 3 (Philadelphia, May 1902): 185–226.

Another North American, the archaeologist Mark Harrington, visited several places in Cuba between 1915 and 1919 and he also made contact with descendants of the Cuban Indigenous people. This investigator leaves for us a very good collection of pictures in which we can observe those somatoscopic features and activities inherited from the aboriginal population. Much of the information procured by him as well as a countless number of archaeological pieces of great value were transferred to United States and are preserved in the National Museum of the American Indian of the Smithsonian Institution, in Washington, DC. Harrington found Indian descendants not only in Santiago de Cuba, but also in the area of La Patana, in Maisí.

In 1952, during a time when many investigators denied their existence, the geneticist Reginald Ruggles Gates demonstrated the presence of these descendants and their blending with white, black, and Chinese people. Gates carried out his studies in the regions of El Caney, Caridad de los Indios, Baracoa, Yara, near Guantánamo, and in Preston. The results can be consulted in the article "Studies in Race Crossing: The Indian Remnants in Eastern Cuba."[115]

In 1964 the departments of Anthropology of the University of Havana and the Academy of Sciences of Cuba co-organized an anthropological expedition to the municipalities of Yateras and Guantánamo, accompanied by foreign scientists from the formerly Soviet socialist republics. The purpose of that work was to conduct a wide study of aboriginal descendants from Cuba in the referred region. Many families already studied by Gates were visited along with other families, also with Indigenous features, in the regions of San Andrés, Monte Verde, Palenquito, Caridad de los Indios, La Escondida, and La Chivera.[116]

Professor Rivero de la Calle first reviewed the results of those early investigations in his important work *The Aboriginal Cultures of Cuba,* published in 1966. He was able to show that these families were made up of individuals who typically present physical features that are different from the other surrounding populations. He concluded that these "correspond with the physical features of the great Mongolic or Asian-

115 Reginald Ruggles Gates, "Studies in race crossing: VI. The Indian remnants in eastern Cuba," *Genetics* 27 (Cambridge, Mass.: Harvard University, 1954): 65–96.

116 Vul'f V. Guinsburg, Milan F. Pospíšil, Manuel Rivero de la Calle, "Expedición antropológica realizada por el Dpto. de Antropología de la Academia de Ciencias y de la Universidad de la Habana," Report, Museo Montané (La Habana: Universidad de la Habana, November 1964).

American race," and that they "differ completely of other mongoloid types arrived to Cuba in later migrations such as the Chinese and the Yucatecan Indians and their descendants who live in the region of Madruga."[117]

In a summary elaborated by the professor himself, with an undefined date but deriving from another study carried out in Yateras in 1973, Rivero de la Calle points out: "We concluded that the population is quite mixed but that no significant differences exist among those denominated as 'pure' and those mixed with Caucasian. We also came to know that the population is very homogeneous since as a group they remained isolated until a relatively recent date.

"The serologic study demonstrated that the population's Indian genes are at about 30 percent. Regardless of this mixture, the morphological type we have considered as Indian offers, in its anthropometric parameters and somatoscopic features, a great similarity with what is known about the American Indian populations in the north of Venezuela. A historical study conducted by the project showed that this population descends in its majority from the old community of El Caney, and that they do not descend from Yucatecan [migrants] as some authors have pointed out."[118]

In 1973, Dr. Rivero de la Calle also elaborated a document, still unpublished, where he explains his work on genealogies carried out with families of aboriginal identifiers in the municipality of Yateras, Caridad de Los Indios, La Sierra, Lagunitas, La Ranchería, Curialitos, Bernardo, La Redonda, Palmar de Güines, Las Coloradas, La Escondida, La Cobrera, San Andrés, Pinar de Ceiba, and others. As part of that investigation, historical studies were conducted on settlement in the region, through interviews with several people. Among them was Antonio Lescaille Ramírez, 91 years old then and residing in Felicidad de Yateras, and Ladislao Ramírez Rojas, who was approximately 110 years old. The team interviewed three of Ladislao's daughters and

117 Manuel Rivero de la Calle, *Las culturas aborígenes de Cuba*, ed. (La Habana: Universitaria, Instituto del Libro, 1966): 57. Madruga, where there is a Maya-descendent community, is in Mayabeque Province, western Cuba.

118 Manuel Rivero de la Calle, "Los indios cubanos de Yateras," *Santiago* 10 (Universidad de Oriente, 1973): 151–174.

consulted court archives and books on vital statistics at Jamaica and Felicidad de Yateras.[119]

Dr. Rivero de la Calle also refers in a report to the results of a trip with anthropological goals conducted in the province of Granma in February 1981. The professor describes locating five families in Yara with "Indian features" and later on, in El Caney de las Mercedes, in the school center Camilo Cienfuegos, where there were almost six thousand children, "...he proved that the areas where the most Indian descendants appear...were: Cienaguilla, in Campechuela and Minas del Frío, Frío de Agua, San Lorenzo, and Jiguaní, although descendants are spread out for almost the whole province, but especially toward the regions in the direction of the Sierra Maestra."[120]

An item pointed out early by Gates is that, among the inhabitants of evident Indigenous origin, even though dispersed through various regions of Guantánamo, two last names prevailed: Ramírez and Rojas. These showed a single lineage dating to the sixteenth century when Manuel Rojas—Diego Velázquez's nephew and *encomendero* (confidant) in the region of Baracoa—held endowments of "Indian slaves." As well, Miguel Ramírez, first bishop of Santiago de Cuba and an appointed "Protector of Indians," celebrated the first massive baptisms in the region, granting their own last names to the newly baptized Indians.121

The Slovak investigator Milan Pospíšil, who studied the referred groups, published in 1976 the book *Indian Remnants from the Oriente Province, Cuba,* where he gives account of the existence of Indigenous methods of cultivating beans and native tubers, the use of slash and burn, as well as some "archaic" tendencies in the Spanish language usage such as calling their chiefs "caciques." He remarks that until recent years they used the mother's last name before the father's, representing a legacy of matrilineal practice.[122] In fact, according to anthropological

119 Manuel Rivero de la Calle, copy of a work report in the personal archive of Dr. Antonio J. Martínez Fuentes.

120 Ibid.

121 Manuel Rivero de la Calle, *Las culturas aborígenes de Cuba*, ed. (La Habana: Universitaria, Instituto del Libro, 1966): 59.

122 Milan F. Pospíšil, *Indian Remnants from the Oriente Province, Cuba*, ed. (Bratislava, Slovakia: Univerzita Komenského, The Nitra Printing Press, Nat Enterprise, 1976): 191.

studies, some early communities were matrilineal until just short of the mid-twentieth century[123].

Without a doubt the growing number of studies in Amerindian populations, especially in the countries that until now had remained unexplored, such as those of the insular Caribbean, are increasingly clear on the complexity of the migration and establishment patterns that in many occasions tend to dim the current vision. The genetic studies have contributed to it in a significant way.

A genetic study carried out by Mendizabal and collaborators, published in 2008, shows that in the current Cuban population, 45 percent of the sequences of the ADNmt haplogroup found came from African origin, 33 percent from Native American origin, and 22 percent from Western Europe. Regarding the haplogroups of the Y chromosome, 79 percent of the analyzed chromosomes can be derived from the genetic component of Western Europe, while the African fraction counts for 20 percent of Cuban lineages, and no masculine Indocuban genetic contribution was found.[124]

The latest genetic investigation conducted in Cuba, carried out by the National Center for Medical Genetics of Cuba and directed by Dr. Beatriz Marcheco Teruel, according to data published in 2014, has proved that on average, 72 percent of the genes in the Cuban current population came from European origin, 20 percent from African origin, and 8 percent from Native American origin. Meanwhile, the study of the mitochondrial DNA and the Y chromosome using uniparental markers showed that in a total of 943 haplotypes, 38.8 percent comes from African origin, 34.5% from Native American, and 26.7 percent from Eurasian. As for the study of the Y chromosome, 81,8 percent comes from Eurasian origin, 17.7 percent from African, and only 0.5 percent from Native American.[125]

123 Daniel Torres Etayo, *Taínos: mitos y realidades de un pueblo sin rostro* (México: Editorial Asesor Pedagógico, 2006).

124 Isabel Mendizabal, Karla Sandoval, Gemma Berniell-Lee, Francesc Calafell, Antonio Salas, Antonio Martínez-Fuentes, David Comas, "Genetic origin, admixture, and asymmetry in maternal and paternal human lineages in Cuba." BMC Evolutionary Biology 8(1) (July 2008): 213. DOI: 10.1186/1471-2148-8-213.

125 Beatriz Marcheco-Teruel, Esteban J Parra, Evelyn Fuentes-Smith, Antonio Salas, Henriette N Buttenschøn, Ditte Demontis, Maria Torres-Español, Lilia C Marin-Padron, Enrique J Gómez-Cabezas, Vanesa Álvarez-Iglesias, Ana Mosquera-Miguel, Antonio Martínez-Fuentes, Ángel Carracedo, Anders D Børglum, Ole Mors,(2014) "Cuba: Exploring the history of admixture and the genetic basis of pigmentation using autosomal and uniparental markers,"Revista PLOSGenetic, http://www.plosgenetics.org/doi/pgen.1004488.

These results confirm the disparity of the maternal and paternal histories of Cuba. The first European immigrations were almost completely masculine. Therefore the biological mixture with aboriginal women—consensual or not—was a necessity for reproduction. Later the intermarriage extended to African males (enslaved and free). The first transculturation of the island was between Spaniards and Amerindians, while the mixture among Africans and Amerindians was very important, blending into marginalized communities such as *palenques* (remote Indian-African villages in resistance). For example, in the seventeenth century, Jiguaní was a refuge for aboriginal people intermarried with Africans who escaped enslavement. Among other Indian pueblos are those founded at Guanabacoa, Tarraco, Ovejas, Guguaney, El Caney (Pichardo Moya, 1945).[126]

III

There is evidence that the cultural trails of these populations survive in Cuba to the present. Many of their expressions, tangible and intangible, are present in our daily life, but we don't recognize their origins. Some scholars affirm that their religious beliefs survive in popular myths and in the syncretism of the African-originated, also called Afro-Cuban religions.[127] According to Pospíšil, many of these communities are Catholic or members of the syncretic religions that exist alongside "healers" and beliefs identified with aboriginal past.[128]

Taíno influence also survives in popular culture. The legend of the *güije*—or *jigüe*—is still told, a small dark creature with a resemblance to a troll. It is said these beings live in rivers and come out to make witchcraft or mischiefs to humans. There are Arawak stories about aquatic deities. Some ascribe these histories to African-originated mythology. Still, "jigüe" resembles the word *jiwe*—"monkey" in continental Arawak language. In the eastern provinces there also exist legends about *cagüeiros*, beings that can transform themselves into animal and plant creatures. All this reminds us of the anthropomorphic

126 Felipe Pichardo Moya, *Los indios de Cuba en sus tiempos historicos* (La Habana: Imprenta El Siglo XX, 1945).

127 Cf. Daisy Fariñas Gutiérrez, *Religión en las Antillas* (La Habana: Editorial Academia, 1995): 150; María Nelsa Trincado, "El aborigen y la Caridad del Cobre," *El Caribe Arqueológico* 2 (Santiago de Cuba, 1997): 114–120.

128 Milan F. Pospíšil, *Indian Remnants from the Oriente Province, Cuba*, ed. (Bratislava, Slovakia: Univerzita Komenského, The Nitra Printing Press, Nat Enterprise, 1976): 225.

transfigurations present in a number of myths and images of Taíno origin.

Jesus Guanche notes in his book *Ethnic Components of the Cuban Nation* that the traditional Cuban soup, the *ajiaco,* is assumed to not include anything that is native Cuban, although ironically the ajiaco is precisely a Taíno dish.[129] The fact is that, when we talk about "cubanity," the aboriginal factor has been mentioned as a side note, without giving importance to its legacy. Don Fernando Ortiz himself recognized the value of the dominant cultures in the Cuban society but practically omitted the Indigenous influence in today's culture. In the text, "Human Factors of Cubanity," published in 1940, he touches on the topic and openly issues his point of view:

> "Cuban culture incorporates very little from Taíno religion, from their gods, cosmogonies, and rites; maybe some superstition on the *siguapa.*
>
> "All the sacrality of the Indians died with them. The idols found today in the caves are lifeless already like the empty skulls of their believers. Everything died but one liturgy and magical ceremony: a sacred rite of the Taínos, discovered here by Christopher Columbus that later, already profaned, was incorporated to the new culture of Cuba and remains as a consuetudinary feature of *cubanity.* Such a rite is the smoking of the tobacco...the earliest, most accepted and most delightful gift given to the universal culture from Cuba.
>
> "In the Indians, mainly in the Taínos who were agrarian, there should already have been present some rudimentary form of *cubanity*, born from the social solidarity of their human group, from their rootedness in their territory, from the cohesive identity of their peculiar culture and from the consciousness of their ancestral unity, even if it's doubtful that some group of the Taíno Indians from Cuba felt with their own historical personality, different from their fellow and progenitors, the Taínos from Haiti. There is no doubt that those Indians felt themselves as Taínos, but it's difficult to assure they felt themselves as Cuban."[130]

---

129 Jesus Guanche, *Componentes étnicos de la nación cubana*, Colección Fuente Viva, Fundación Fernando Ortiz (La Habana: Ediciones Unión, 2008).

130 Fernando Ortiz, "Los factores humanos de la cubanidad," *Etnia y sociedad* (La Habana: Editorial de Ciencias Sociales, 1993).

In this important work, Ortiz enforces his idea of a Cuban aboriginal culture gone extinct. Of this idea Rivero de la Calle remarked:

"We believe that one of the causes that motivated certain skepticism about the survival of possible descendants from our aboriginal people in our country was the lack of knowledge about an entire series of anthropological investigations carried out from 1952 and started on that date by the English geneticist R. R. Gates, and the poor diffusion they have received in Cuba. Additionally the polygraph Don Fernando Ortiz always stood against the idea of Indigenous survival and naturally, his enormous scientific prestige and influence exercised for many decades over the anthropological sciences cannot be denied."[131]

More recently Robaina Jaramillo, Celaya González, and Pereira Pereira emphasized:

> "This Ortizian approach drove, at least in the scientific cultural environment, the belief in the total disappearance of the Indian [people] and it strengthened the supposition they had been completely exterminated in Cuba since the sixteenth century. In his definitive way...Ortiz saw [the Indian] as dead, with no strength in the social breath of the country...despite the Indigenous material culture, deeply studied by him and which he connected to our history, to the genesis of the national process, let's say again, in terms of continuity and historical presence.
>
> "Perhaps the great prestige of a personality as Ortiz's has determined the prevalence of his statement over time. Today, the idea of the 'extinction' of Indians and their culture still prevail. Certainly, Ortiz described the important components provided by the African population, mainly as incorporated to the Spanish trunk of Cuban culture, but excluded the native components. And so, his vision of cubanity in this way limits the live ethnic and cultural elements of the society to the Spanish and the African, with the unimportant addition of other vestigial components that arrived to Cuba through the constant immigrations from the most diverse corners on the planet.

131 Manuel Rivero de la Calle, "Los indios cubanos de Yateras," *Santiago* 10 (Universidad de Oriente, 1973): 151–174.

> "Unfortunately, Ortiz himself omitted in his concept of cubanity and for our self-concept in today's cultural identity, the first transculturation process in the genesis of the Cuban nation, the one that necessarily took place between Indian and Spanish, between Indian and African."[132]

What happened to the aboriginal population of Cuba was a horrible thing, but there were survivors who were able to pass their culture, alongside with their genes, up until our days. I agree with Valdés Bernal when he remarks that:

> "...the bloody and imposed transculturation process, of biological and cultural mixture, gave birth to a *criollo* population of Indigenous roots which constituted the first substratum of our nation. Therefore, what really happened was a demographic ethnocide, not a cultural genocide, the 'Indian' component was preserved in our *ajiaco* [mess] in an entire series of cultural expressions and in our national language."[133]

The unproblematized Caribbean that some insist on seeing as simply the place where cultures coming from three continents first met, was in fact the scenario where the biggest holocaust in our history started, arguably the biggest ethnic cleansing of all time . Those who seek to erase that past are doing too little for themselves, and are forgetting that the past implies that those cultural and biological principles are part of what today we know as the Cuban Nation. We are the result of an amalgam that began more than 500 years ago, which includes the contribution of the American aboriginal, which is our first root.

To know who we are and where we came from evidently helps us make better sense of our lives. Today in different countries of the continent and all over the world we can find diverse processes that seek to value Indigenous identity—from the rescue of native languages, medical practices, artistic expressions, orality, and more—and Cuba should not be the exception. Even if we don't contemplate the existence of an ethnic Indigenous minority as such, it is important to recognize

132 Jesús Rafael Robaina Jaramillo, Miriam Celaya González, Oscar Pereira Pereira, "La arqueología en la construcción de un discurso sobre identidad cultural cubana, *Catauro* 8 (2003): 47–61.

133 Sergio Valdés Bernal, "¿Y nuestro abuelo indio dónde está?" *Antropológicas,* no 36 (2012), available at http://intranet.uh.cu/servicios-de-informacion/publicaciones-de-nuestra-universidad/antropologicas-publicacion-semanal.

that a legacy does exist and cannot continue to be ignored or banished to a distant time. If cubanity is the consciousness and the will of being Cuban, it is fundamental to identify and to recognize all the cultures that constitute it.

IV

My first approach to Panchito's world begins after qualifying for the Biological Sciences degree at the University of Havana (UH) in September 1964, where two years later Dr. Rivero de la Calle, head of the Department of Anthropology at the time, accepted me as his student assistant to work under his guidance. Data obtained in fieldwork was still processed by that time. The professor Milan Pospíšil was still in the Department of Anthropology of the Biological Sciences School of the UH and Professor Rivero gave me a task to support him in the processing of the obtained information, an occasionally tedious work that I had to include among my student duties.

There were several months of work, and I'm sure through my hand passed the data of the then-young Panchito Ramírez . And likely so, because I haven't forgotten the professor's stories about Ladislao Ramírez, Panchito's grandfather, and Marcelina Ramírez, one of his cousins. In the book *The Aboriginal Cultures of Cuba*, written by Rivero de la Calle and published in 1966, is a photo of Milan Pospíšil with Marcelina at Caridad de los Indios.

At the beginning of the 1970s, I had the opportunity to travel to several places in the region of Yateras while collaborating with a national investigation on the growth and development of Cuban children and youth, directed by the then Institute of Childhood. I was near La Ranchería, but logistical issues prevented me from getting there.

Now, thanks to this book, I have met again with my own memoirs and experiences in that region and I can do it through the simple yet profound cosmic vision of Panchito—his values, ethics, respect, spirituality, the jigüe. Here I find his inherited knowledge on the corn, the yuca, the tobacco, the beans, the value of the *conuco* (field), the wood, the natural medicine, the water, the sun, and the moon. Panchito speaks to scientists in his own language about aspects such as the climate change, the drought, the sowing; he requests that scientists investigate what is happening, and he asks us: "Why are these changes happening, changes never seen by this *guajiro*'s eyes? Figure out," he says, "what is happening with nature."

Panchito's voice narrates and at the same time questions us, as he is aware and vindicates from his Indigenous origins. In his hills, the people live proud of their ancestors, their customs, and their traditions, and they are aware of the necessity to take care and to love Mother Earth. They know they palpitate in the center of a history that makes them unique, full with beautiful legends and experiences we might study and spread out systematically. He is as much Cuban as he is Indian, and he surprises us when he tells us:
"I pray to and love Martí because Martí covers the sky of all Cubans. Martí covers the four Cuban directions, and from the Earth up to the sky.

> "And to Earth. I pray to the Earth, too. Because beyond what we do as men, She still is. There are so many things I pray to Earth. And so many things that the Earth tells me." [134]

Precisely, it is José Martí who reminds us:

> "The Indigenous bring us a new way of life. We study what was brought to us from France; but they will reveal to us what they take from Nature. From copper faces will shine a new light. The teachings will reveal us to ourselves ."[135]

I perceive that light, that teaching and wisdom, that ecological knowledge, in every word Panchito has told to Barreiro—a wisdom not learned in books but in direct contact with Mother Earth. I perceive that light in Panchito's bond with all that he inherited, also in the teachings of Martí. As expressed by the new generation colleagues from the Kaweiro Group: "Panchito and some others in La Ranchería preserve a certain bond with what could be called the silent knowledge: dreams, visions, conversations with plants and animals and other forms of being and doing that keep their lives connected to the mystery and the marvel of being human, that perception that the city noise does not allow without interference and a noise that, regretfully, the new generations of the

---

134 José Barreiro, *Panchito, Mountain Cacique: A Guajiro-Taíno Testimony from Francisco Ramírez Rojas* (Santiago de Cuba: Catedral Ediciones, 2001).

135 José Martí, J, El proyecto de Instruccion publica,".- Los Articulos de la Fe.-La Enseñanza Obligatoria, En Obras Completas, t. 2, Edición Critica. La Habana: Centro de Estudios Martianos, 2006, p. 212.

community begin to be experience. Farther away from the common core of the community for several reasons, many now live in the cities with more or less degree of consciousness of their origins. Even though La Ranchería is another example of firmness against the breakneck speed of the Eurocentric culture (of which Cuba is part of), which has lost not a piece of its aggressiveness toward 'the other' and looks for, now more subtly disguised behind the veil of 'development,' the homogenization of a worldview."[136]

In the twenty-first century, the voice of Panchito represents for us the continuity of those millennial cultures that were kidnapped, whisked away from our identity. He is Indian, he feels like an Indian and the Indian speaks through his mouth here and now. Which is why to the question of how it can be possible, after so many years, that Indigenous descendants continue to exist in Cuba, he answers in the old philosophy of his lineage: "You sow a tree and that tree finally dries out, but it leaves seeds that grow; that's why the Indian never ends." [137]

How would anyone dare to deny a man who speaks that way—how to deny him and the many other Panchitos of our archipelago—the right to identify himself or herself as a member of that supposedly extinct culture? No one can assume the right to erase the history of our origins. No one can limit us by removing the richness and diversity that makes the present possible, that will allow the future. Consequently, we must thank Panchito for his words, for the wisdom he shares day to day with the youth of the community where he was born and where he is honored as *cacique* (chief). And, also, let us give our thanks to José Barreiro, whose commitment to Cuba and to the Taíno roots, has allowed him to amplify and to share Panchito's voice, so it can take its place—as it should—in the fold of our inheritance.

Antonio J. Martínez Fuentes

A director of the Montané Anthropological Museum for 35 years, and scholar emeritus in the Department of Human and Animal Biology, University of Havana, Dr. Antonio J. Martinez is a widely recognized Cuban scientist.

136 Recorrido del Grupo Kaweiro por el Oriente de Cuba: En La Ranchería (Guantánamo), published March 20, 2013, http://www.grupokweiro.netai.net/index.php/22-art-gira-oriental-2012-rancheria-es.

137 Yisell Rodriguez Milán, "Los Indios de la Caridad," accessed May 6, 2013, https://lentedeaumento.wordpress.comtag|a-caridad.

About the Author

José Barreiro is an author and intellectual activist with forty years dedicated service to Native peoples and issues. Presently, Smithsonian Scholar Emeritus, Barreiro retired after a decade as assistant director for research and director, Office for Latin America, Smithsonian National Museum of the American Indian, in December 2016. Previously senior editor and chief editorialist of Indian Country Today, Barreiro is one of the nation's leading voices in American Indian policy, journalism, and publishing. For 20 years he helped forge the American Indian Program at Cornell University, where he supervised dozens of research projects by historians, anthropologists, geographers, and multi-disciplinary scholars to produce numerous forums and edit nearly one hundred scholarly articles and several books from among the country's leading authors.

A member of the Taíno Nation of the Antilles, Barreiro has focused on Indigenous themes of the Caribbean, both on contemporary and historical topics. Barreiro's articles in this area regularly appear in national anthologies such as "The Cuba Reader" (Duke University Press), "Wild Majesty" (Oxford), and «Caribbean Indigenous Resurgence» (Peter Lang, Inc.). Jose Barreiro holds a Ph.D. in American studies from the State University of New York at Buffalo. He is married to Katsi Cook (Mohawk), traditionalist midwife and reproductive health advocate. They have six children and eleven grandchildren.

Gulf of Mexico
La Habana
Artemisa
Mayabeque
Matanzas
Pinar del Río
Cienfuegos
Isla de la Juventud
Caribbean Sea

ATLANTIC OCEAN
Ciego
de Ávila
Camagüey
Las Tunas
Holguín
Granma
Santiago
de Cuba
Guantánamo

www.ingramcontent.com/pod-product-compliance
Lightning Source LLC
Chambersburg PA
CBHW050230160826
49840CB00003B/132
9781962756044

"José Barreiro's work with Cuban Indian cacique Panchito reaffirms a Native cultural continuity for Cuba that has been largely ignored. Dreaming Mother Earth is a labor of love. Over a lifetime of love and labor, Barreiro has chronicled many Native people and provided a map to the spiritual and land based constellations of our ways of life.

Barreiro remains a lifetime storyteller, impeccable scholar, and journalist—from his early days at the Native movement publication, Akwesasne Notes, in the 1970s, to this coolest of books, an actual Native chronicle from contemporary Cuba. I am forever grateful for his love of our people, history, and research."

–Winona LaDuke, Native author and scholar-activist

"Thanks for your 'Panchito'... a fountain of *valiosísima* information ... how much I would have liked to be with you in that journey to the mountain."

– Dr. José Juan Arrom, (letter to the author)
* Dr. Jose Juan Arrom (Yale emeritus) d. 2008) is among the most distinguished Caribbean scholars of the 20th Century

"... the most extensive and significant recent work is that by José Barreiro; [it has] served to bring the issue of Indigenous survival in Cuba back onto an international agenda. ... Refreshingly, Barreiro is not much interested in physical appearance. Rather, he spoke to people and listened to what they had to say... Unlike Culin, Barreiro is not looking for lost Indians. Rather, he embraces both the process of transculturation which has embedded Panchito and his community into the sierra and the signs of the survival of Indigenous practices which provide the community with a sense of identity ..."

– Peter Hulme, Oxford University,
author of *Cuba's Wild East: A Literary Geography of Oriente* (American Tropics), Liverpool University Press (2011)

"Barreiro [is] challenging the essentialist notions of culture that would sequester Amerindians to an ancient past and thereby deny that the much-vaunted hybridity of the modern Caribbean includes Indigenous hybridity as well."

– Maxemilian C. Forte, Concordia University,
author of "Extinction: Ideologies Against Indigeneity in the Caribbean" Southern Quarterly, Summer 2006

Gulf of Mexico
La Habana
Artemisa
Mayabeque
Matanzas
Villa Clara
Pinar del Río
Cienfuegos
Isla de la Juventud
Caribbean Sea